Essential Student Algebra

VOLUME FOUR

Linear Algebra

*Essential
Student
Algebra*

VOLUME FOUR

Linear Algebra

T. S. BLYTH & E. F. ROBERTSON
University of St Andrews

London   New York
CHAPMAN AND HALL

First published in 1986 by
Chapman and Hall Ltd
11 New Fetter Lane, London EC4P 4EE
Published in the USA by
Chapman and Hall
29 West 35th Street, New York NY 10001

© 1986 T. S. Blyth and E. F. Robertson

Printed in Great Britain by
J. W. Arrowsmith Ltd., Bristol

ISBN 0 412 27850 2

This paperback edition is sold subject to the condition that it shall not, by way of trade or otherwise, be lent, resold, hired out, or otherwise circulated without the publisher's prior consent in any form of binding or cover other than that in which it is published and without a similar condition including this condition being imposed on the subsequent purchaser.

All rights reserved. No part of this book may be reprinted or reproduced, or utilized in any form or by any electronic, mechanical or other means, now known or hereafter invented, including photocopying and recording, or in any information storage and retrieval system, without permission in writing from the publisher.

British Library Cataloguing in Publication Data
Blyth, T. S.
Essential student algebra.
Vol 4: Linear algebra
1. Algebra
I. Title   II. Robertson, E. F.
    512   QA155
ISBN 0-412-27850-2

131360

BELMONT COLLEGE LIBRARY

Contents

Preface

Chapter One : *The minimum polynomial*	1
Chapter Two : *Direct sums of subspaces*	7
Chapter Three : *Reduction to triangular form*	22
Chapter Four : *Reduction to Jordan form*	31
Chapter Five : *The rational and classical forms*	44
Chapter Six : *Dual spaces*	58
Chapter Seven : *Inner product spaces*	69
Chapter Eight : *Orthogonal direct sums*	87
Chapter Nine : *Bilinear and quadratic forms*	99
Chapter Ten : *Real normality*	108
Index	119

Preface

If, as it is often said, mathematics is the queen of science then algebra is surely the jewel in her crown. In the course of its vast development over the last half-century, algebra has emerged as the subject in which one can observe pure mathematical reasoning at its best. Its elegance is matched only by the ever-increasing number of its applications to an extraordinarily wide range of topics in areas other than 'pure' mathematics.

Here our objective is to present, in the form of a series of five concise volumes, the fundamentals of the subject. Broadly speaking, we have covered in all the now traditional syllabus that is found in first and second year university courses, as well as some third year material. Further study would be at the level of 'honours options'. The reasoning that lies behind this modular presentation is simple, namely to allow the student (be he a mathematician or not) to read the subject in a way that is more appropriate to the length, content, and extent, of the various courses he has to take.

Although we have taken great pains to include a wide selection of illustrative examples, we have not included any exercises. For a suitable companion collection of worked examples, we would refer the reader to our series *Algebra through practice* (Cambridge University Press), the first five books of which are appropriate to the material covered here.

T.S.B., E.F.R.

CHAPTER ONE

The minimum polynomial

In Volume Two we introduced the notions of *eigenvalue* and *eigenvector* of a linear mapping or matrix. There we concentrated our attention on showing the importance of these notions in solving particular problems. Here we begin by taking a closer algebraic look.

Definition Let F be a field. By an *algebra* over F we shall mean a vector space V over F on which there is defined a multiplication in such a way that $(V, +, \cdot)$ is a ring with identity and

$$(\forall x, y \in V)(\forall \lambda \in F) \qquad (\lambda x)y = \lambda(xy) = x(\lambda y).$$

Example With respect to multiplication of matrices, the vector space $\text{Mat}_{n \times n}(F)$ becomes an algebra.

Example With respect to composition of mappings, the vector space $\text{Lin}(V, V)$ becomes an algebra.

If V is a vector space of dimension n over F then we have an algebra isomorphism

$$\text{Lin}(V, V) \simeq \text{Mat}_{n \times n}(F)$$

(i.e. a bijection that is both a ring and a vector space isomorphism) that is obtained by associating with each linear mapping $f : V \to V$ its matrix relative to some fixed ordered basis. This is well known; see, for example, Theorems 7.2 and 7.3 in Volume Two. In practice, we work in both of these algebras, choosing whichever suits our purposes at the time. Observe, for example, that since $\text{Mat}_{n \times n}(F)$ is of dimension n^2 over F, for every $n \times n$ matrix A over F the $n^2 + 1$ powers

$$I_n = A^0, A, A^2, A^3, \ldots, A^{n^2}$$

are linearly dependent and so there is a non-zero polynomial

$$p = a_0 + a_1 X + a_2 X^2 + \cdots + a_{n^2} X^{n^2} \in F[X]$$

such that $p(A) = 0$. The same of course is true for any $f \in \text{Lin}(V, V)$. But we can do better than this : there is, in fact, a polynomial p *of degree at most* n such that $p(A) = 0$. This is the celebrated *Cayley-Hamilton Theorem* which we shall now establish. Since we shall be working in $\text{Mat}_{n \times n}(F)$, the proof we shall give will be 'elementary'. There are other, more elegant, proofs which use $\text{Lin}(V, V)$.

Definition If $A \in \text{Mat}_{n \times n}(F)$ the the *characteristic polynomial* of A is

$$\chi_A = \det(X I_n - A).$$

Note that χ_A is of degree n in the indeterminate X.

1.1 Theorem [Cayley-Hamilton] $\chi_A(A) = 0$.

Proof Let $B = X I_n - A$ and

$$\chi_A = \det B = b_0 + b_1 X + \cdots + b_n X^n.$$

Consider the matrix $\text{adj}\, B$. By definition, this is an $n \times n$ matrix whose entries are polynomials in X of degree at most $n - 1$ and so we have

$$\text{adj}\, B = B_0 + B_1 X + \cdots + B_{n-1} X^{n-1}$$

for some $n \times n$ matrices $B_0, \ldots, B_{n-1}$. Recalling that $B \text{adj}\, B = (\det B) I_n$ (see, for example, Theorem 8.11 of Volume Two), we have

$$(\det B) I_n = B \text{adj}\, B = (X I_n - A) \text{adj}\, B$$
$$= X \text{adj}\, B - A \text{adj}\, B,$$

i.e. we have the polynomial identity

$$b_0 I_n + b_1 I_n X + \cdots + b_n I_n X^n$$
$$= B_0 X + \cdots + B_{n-1} X^n - A B_0 - \cdots - A B_{n-1} X^{n-1}.$$

Equating coefficients of like powers, we obtain

$$b_0 I_n = -AB_0$$
$$b_1 I_n = B_0 - AB_1$$
$$\vdots$$
$$b_{n-1} I_n = B_{n-2} - AB_{n-1}$$
$$b_n I_n = B_{n-1}.$$

Multiplying the first equation on the left by I_n, the second by A, the third by A^2, and so on, we obtain

$$b_0 I_n = -AB_0$$
$$b_1 A = AB_0 - A^2 B_1$$
$$\vdots$$
$$b_{n-1} A^{n-1} = A^{n-1} B_{n-2} - A^n B_{n-1}$$
$$b_n A^n = A^n B_{n-1}.$$

Adding these equations together, we obtain $\chi_A(A) = 0$. $\diamond$

The Cayley-Hamilton Theorem is really quite remarkable, it being far from obvious that an $n \times n$ matrix over a field F should satisfy a polynomial equation of degree n. This result leads us to consider the following notion.

Definition If $A \in \text{Mat}_{n \times n}(F)$ then the *minimum polynomial* of A is the monic polynomial m_A of least degree such that $m_A(A) = 0$.

1.2 Theorem *If p is a polynomial such that $p(A) = 0$ then the minimum polynomial m_A divides p.*

Proof By euclidean division there are polynomials q, r such that $p = m_A q + r$ with $r = 0$ or $\deg r < \deg m_A$. Now by hypothesis $p(A) = 0$, and by definition $m_A(A) = 0$. Consequently we have $r(A) = 0$. By the definition of m_A we cannot then have $\deg r < \deg m_A$, and so we must have $r = 0$. It follows that $p = m_A q$ and so m_A divides p. $\diamond$

1.3 Corollary m_A *divides* χ_A. $\diamond$

It is immediate from 1.3 that every zero of m_A is a zero of χ_A. The converse is also true :

1.4 Theorem m_A and χ_A have the same zeros.

Proof Observe that if λ is a zero of χ_A then $\det(\lambda I_n - A) = 0$ and so $\lambda I_n - A$ is not invertible. There is therefore a dependence relation between the columns of $\lambda I_n - A$ and so there is a non-zero $\mathbf{x} \in \text{Mat}_{n \times 1}(F)$ such that $A\mathbf{x} = \lambda \mathbf{x}$. Given any
$$h = a_0 + a_1 X + \cdots + a_k X^k$$
we then have
$$\begin{aligned} h(A)\mathbf{x} &= a_0 \mathbf{x} + a_1 A \mathbf{x} + \cdots + a_k A^k \mathbf{x} \\ &= a_0 \mathbf{x} + a_1 \lambda \mathbf{x} + \cdots + a_k \lambda^k \mathbf{x} \\ &= h(\lambda) \mathbf{x} \end{aligned}$$
so that $h(\lambda)I_n - h(A)$ is not invertible. Put another way, we have $\det[h(\lambda)I_n - h(A)] = 0$. Thus we see that $h(\lambda)$ is a zero of $\chi_{h(A)}$. Now choose $h = m_A$. Then for every zero λ of χ_A we have that $m_A(\lambda)$ is a zero of
$$\chi_{m_A(A)} = \chi_0 = \det X I_n = X^n.$$
Since the only zeros of this are 0, we have $m_A(\lambda) = 0$ and so λ is a zero of m_A. $\diamond$

Example The characteristic polynomial of
$$A = \begin{bmatrix} 1 & 0 & 1 \\ 0 & 2 & 1 \\ -1 & 0 & 3 \end{bmatrix}$$
is $\chi_A = (X-2)^3$. Now it is readily seen that $A - 2I_3 \neq 0$ and $(A - 2I_3)^2 \neq 0$, so we also have $m_A = (X-2)^3$.

Example For the matrix
$$A = \begin{bmatrix} 5 & -6 & -6 \\ -1 & 4 & 2 \\ 3 & -6 & -4 \end{bmatrix}$$
we have $\chi_A = (X-1)(X-2)^2$. By 1.4, the minimum polynomial is therefore either $(X-1)(X-2)^2$ or $(X-1)(X-2)$. Since, as is readily seen, $(A-I_3)(A-2I_3) = 0$, it follows that $m_A = (X-1)(X-2)$.

The notion of characteristic polynomial can be defined for a linear mapping as follows. Given a vector space V of dimension n over F and a linear mapping $f : V \to V$, let A be the matrix of f relative to some fixed ordered basis of V. Then the matrix of f relative to any other ordered basis is of the form $P^{-1}AP$ where P is the transition matrix from the new basis to the old basis (see, for example, Volume Two, Theorem 7.6). Now the characteristic polynomial of $P^{-1}AP$ is

$$\begin{aligned}\det(XI_n - P^{-1}AP) &= \det[P^{-1}(XI_n - A)P] \\ &= \det P^{-1} \det(XI_n - A) \det P \\ &= \det(XI_n - A),\end{aligned}$$

i.e. $\chi_{P^{-1}AP} = \chi_A$. Thus the characteristic polynomial is independent of the choice of basis, so we can define the characteristic polynomial χ_f of f to be the characteristic polynomial of any matrix that represents f. Likewise, the minimum polynomial m_f of f is defined to be the minimum polynomial of any matrix that represents f; for if A, B represent f then for any polynomial p we have $p(B) = p(P^{-1}AP) = P^{-1}p(A)P$ and so $p(B) = 0$ if and only if $p(A) = 0$.

As we have seen, the characteristic polynomial and the minimum polynomial have the same zeros. These are called the *eigenvalues* (of f or of A). Thus λ is an eigenvalue of A if and only if $\det(\lambda I_n - A) = 0$, and the corresponding statement for f is that λ is an eigenvalue of f if and only if $\lambda \operatorname{id}_V - f$ is not invertible. In the former case there exists a non-zero $\mathbf{x} \in \operatorname{Mat}_{n \times 1}(F)$ such that $A\mathbf{x} = \lambda \mathbf{x}$, and in the latter there exists a non-zero $x \in \operatorname{Ker}(\lambda \operatorname{id}_V - f)$, so that $f(x) = \lambda x$. Such a column matrix $\mathbf{x}$ and vector x are called corresponding *eigenvectors* (of A and of f).

1.5 Theorem *Let V be a vector space of dimension $n \geq 1$ over $\mathbb{C}$. Then every $f \in \operatorname{Lin}(V, V)$ and every $A \in \operatorname{Mat}_{n \times n}(\mathbb{C})$ has at least one eigenvalue in $\mathbb{C}$.*

Proof χ_f factorises over $\mathbb{C}$, say as

$$\chi_f = (X - \lambda_1)^{e_1}(X - \lambda_2)^{e_2} \cdots (X - \lambda_k)^{e_k}.$$

Substituting f for X and using the Cayley-Hamilton Theorem, we obtain

$$0 = (f - \lambda_1 \operatorname{id}_V)^{e_1}(f - \lambda_2 \operatorname{id}_V)^{e_2} \cdots (f - \lambda_k \operatorname{id}_V)^{e_k}.$$

It follows that not all the factors $(f - \lambda_i \, \mathrm{id}_V)^{e_i}$ are invertible and so there is at least one eigenvalue. ◇

Example Note that 1.5 is not true when $\mathbb{C}$ is replaced by $\mathbb{R}$. Indeed, consider the rotation matrix

$$R_\vartheta = \begin{bmatrix} \cos \vartheta & \sin \vartheta \\ -\sin \vartheta & \cos \vartheta \end{bmatrix}.$$

The characteristic polynomial of R_ϑ is $X^2 - 2\cos\vartheta \, X + 1$, the zeros of which are $\cos\vartheta \pm i \sin\vartheta$. Thus, when ϑ is not an integral multiple of π, R_ϑ has no real eigenvalues.

1.6 Theorem *A linear mapping f (or a square matrix A) is invertible if and only if the constant term in the characteristic polynomial is not zero.*

Proof To say that f is invertible is equivalent to saying that 0 is not an eigenvalue of f, i.e. to saying that 0 is not a zero of the characteristic polynomial. Clearly, this is equivalent to the constant term being non-zero. ◇

Example The matrix

$$A = \begin{bmatrix} 1 & 1 & 1 \\ 0 & 1 & 1 \\ 0 & 0 & 1 \end{bmatrix}$$

is such that $\chi_A = (X-1)^3$. Thus

$$0 = (A - I_3)^3 = A^3 - 3A^2 + 3A - I_3$$

and consequently we see that

$$A^{-1} = A^2 - 3A + 3I_3 = \begin{bmatrix} 1 & -1 & 0 \\ 0 & 1 & -1 \\ 0 & 0 & 1 \end{bmatrix}.$$

CHAPTER TWO

Direct sums of subspaces

If A and B are non-empty subsets of a vector space V over a field F then the subspace spanned by $A \cup B$, i.e. the smallest subspace that contains both A and B, is the set of linear combinations of elements of $A \cup B$. In other words, it is the set of elements of the form $\sum_{i=1}^{m} \lambda_i x_i + \sum_{j=1}^{n} \mu_j y_j$ where each $x_i \in A$, each $y_j \in B$, and $\lambda_i, \mu_j \in F$. In the case where A, B are *subspaces* of V, this set can be described as

$$A + B = \{a + b \,;\, a \in A, b \in B\}$$

which we call the *sum* of the subspaces A, B. More generally, if $A_1, \ldots, A_n$ are subspaces of V then we define their *sum* to be the subspace, denoted by $\sum_{i=1}^{n} A_i$, that is spanned by $\bigcup_{i=1}^{n} A_i$. Clearly, we have

$$\sum_{i=1}^{n} A_i = \{a_1 + \cdots + a_n \,;\, a_i \in A_i\}.$$

Example Let X, Y, D be the subspaces of $\mathbb{R}^2$ given by

$$X = \{(x, 0) \,;\, x \in \mathbb{R}\}, \quad Y = \{(0, y) \,;\, y \in \mathbb{R}\},$$
$$D = \{(x, x) \,;\, x \in \mathbb{R}\}.$$

Then $\mathbb{R}^2 = X + Y = X + D = Y + D$, for every $(x, y) \in \mathbb{R}^2$ can be written in each of the three ways

$$(x, 0) + (0, y), \quad (x - y, 0) + (y, y), \quad (0, y - x) + (x, x).$$

Definition A sum $\sum_{i=1}^{n} A_i$ of subspaces $A_1, \ldots, A_n$ is said to be *direct* if every $x \in \sum_{i=1}^{n} A_i$ can be written in a unique way as a sum $x = a_1 + \cdots + a_n$ with $a_i \in A_i$ for each i.

We shall use the notation $\bigoplus_{i=1}^{n} A_i$ to denote the fact that the sum $\sum_{i=1}^{n} A_i$ is direct, and call this the *direct sum* of the subspaces $A_1, \ldots, A_n$.

Example In the previous Example we have $\mathbb{R}^2 = X \oplus Y = X \oplus D = Y \oplus D$.

Example Let A, B be the subspaces of $\mathbb{R}^3$ given by

$$A = \{(x, y, z) \; ; \; x + y + z = 0\}, \quad B = \{(x, x, z) \; ; \; x, z \in \mathbb{R}\}.$$

Then $\mathbb{R}^3 = A + B$ since, for example,

$$(x, y, z) = \left(\tfrac{1}{2}(x-y), -\tfrac{1}{2}(x-y), 0\right) + \left(\tfrac{1}{2}(x+y), \tfrac{1}{2}(x+y), z\right).$$

This sum is not direct, however, for we can also write (x, y, z) as

$$\left(\tfrac{1}{2}(x-y+1), -\tfrac{1}{2}(x-y-1), -1\right) + \left(\tfrac{1}{2}(x+y-1), \tfrac{1}{2}(x+y-1), z+1\right).$$

2.1 Theorem *If $A_1, \ldots, A_n$ are subspaces of a vector space V then the following statements are equivalent :*

(1) *the sum $\sum_{i=1}^{n} A_i$ is direct;*

(2) *if $\sum_{i=1}^{n} a_i = 0$ with $a_i \in A_i$ for every i, then every $a_i = 0$;*

(3) *for every i, $A_i \cap \sum_{j \neq i} A_j = \{0_V\}$.*

Proof $(1) \Rightarrow (2)$: By the definition of direct sum, if (1) holds then 0_V can be written in only one way as a sum $\sum_{i=1}^{n} a_i$ with $a_i \in A_i$ for every i.

$(2) \Rightarrow (3)$: Let $x \in A_i \cap \sum_{j \neq i} A_j$, say $x = a_i = \sum_{j \neq i} a_j$. We can write this as $a_i - \sum_{j \neq i} a_j = 0$. By (2) we deduce that $a_i = 0$, whence $x = 0$.

(3) ⇒ (1) : Suppose that (3) holds and that $\sum_{i=1}^{n} a_i = \sum_{i=1}^{n} b_i$ where $a_i, b_i \in A_i$ for each i. Then

$$a_i - b_i = \sum_{j \neq i} (b_j - a_j)$$

where the left hand side belongs to A_i and the right hand side belongs to $\sum_{j \neq i} A_j$. By (3) we deduce that $a_i - b_i = 0$. Since this holds for every i, (1) follows. ◇

2.2 Corollary *If A, B are subspaces of V then $V = A \oplus B$ if and only if $V = A + B$ and $A \cap B = \{0_V\}$.* ◇

Example A mapping $f : \mathbb{R} \to \mathbb{R}$ is said to be *even* if $f(-x) = f(x)$ for every $x \in \mathbb{R}$, and *odd* if $f(-x) = -f(x)$ for every $x \in \mathbb{R}$. The sets A, B of even, odd functions are subspaces of the vector space $V = \text{Map}(\mathbb{R}, \mathbb{R})$. Moreover, $V = A \oplus B$. To see this, given any $f : \mathbb{R} \to \mathbb{R}$ let $f^+ : \mathbb{R} \to \mathbb{R}$ and $f^- : \mathbb{R} \to \mathbb{R}$ be given by $f^+(x) = \frac{1}{2}[f(x) + f(-x)]$ and $f^-(x) = \frac{1}{2}[f(x) - f(-x)]$. Then f^+ is even and f^- is odd. Since $f = f^+ + f^-$ we have $V = A + B$. Since clearly $A \cap B$ consists only of the zero function, it follows by 2.2 that $V = A \oplus B$.

Example Let V be the vector space $\text{Mat}_{n \times n}(\mathbb{R})$. If A, B are the subspaces of V consisting of the symmetric, skew-symmetric matrices then $V = A \oplus B$. In fact, every $X \in V$ can be written uniquely in the form $X = Y + Z$ where $Y \in A$ and $Z \in B$; we have $Y = \frac{1}{2}(X + X^t)$ and $Z = \frac{1}{2}(X - X^t)$.

In a direct sum, bases can be pasted together :

2.3 Theorem *Let V be a finite-dimensional vector space and let $V_1, \ldots, V_n$ be non-zero subspaces of V such that $V = \bigoplus_{i=1}^{n} V_i$. If B_i is a basis of V_i for each i then $\bigcup_{i=1}^{n} B_i$ is a basis of V.*

Proof Let $\dim V_i = d_i$ and let $B_i = \{e_{i,1}, \ldots, e_{i,d_i}\}$. Since $V = \bigoplus_{i=1}^{n} V_i$ we have $V_i \cap \sum_{j \neq i} V_j = \{0_V\}$ by 2.1 and hence $V_i \cap V_j = \{0_V\}$ for $i \neq j$. Consequently $B_i \cap B_j = \emptyset$ for $i \neq j$. Now a typical element of the subspace spanned by $\bigcup_{i=1}^{n} B_i$ is of the form

$$(1) \qquad \sum_{j=1}^{d_1} \lambda_{1j} e_{1,j} + \cdots + \sum_{j=1}^{d_n} \lambda_{nj} e_{n,j}$$

i.e. of the form

$$(2) \qquad x_1 + \cdots + x_n \quad \text{where} \quad x_i = \sum_{j=1}^{d_i} \lambda_{ij} e_{i,j}.$$

Since $V = \sum_{i=1}^{n} V_i$ and since B_i is a basis of V_i it is clear that every $x \in V$ can be expressed in the form (1) and so V is spanned by $\bigcup_{i=1}^{n} B_i$. If now in (2) we have $x_1 + \cdots + x_n = 0_V$ then by 2.1 we deduce that each $x_i = 0_V$ and consequently each $\lambda_{ij} = 0$. Thus $\bigcup_{i=1}^{n} B_i$ is a basis of V. $\diamond$

2.4 Corollary $\dim \bigoplus_{i=1}^{n} V_i = \sum_{i=1}^{n} \dim V_i$. $\diamond$

We shall now determine precisely when a vector space is a direct sum of finitely many non-zero subspaces. As we shall see, this is closely related to the following types of linear mapping.

Definition Let A, B be subspaces of a vector space V such that $V = A \oplus B$, so that every $x \in V$ can be expressed uniquely in the form $x = a + b$ where $a \in A$ and $b \in B$. By the *projection on A parallel to B* we mean the linear mapping $p : V \to V$ given by $p(x) = a$.

Example We know that $\mathbb{R}^2 = X \oplus D$ where $X = \{(x, 0) \; ; \; x \in \mathbb{R}\}$ and $D = \{(x, x) \; ; \; x \in \mathbb{R}\}$. The projection on X parallel to D is given by $p(x, y) = (x - y, 0)$. Thus the image of the point (x, y) is the point of intersection with X of the line through (x, y) parallel to to the line D. The terminology used is thus suggested by the geometry.

Definition A linear mapping $f : V \to V$ is said to be a *projection* if there are subspaces A, B such that $V = A \oplus B$ and f is the projection on A parallel to B. A linear mapping $f : V \to V$ is said to be *idempotent* if $f \circ f = f$.

2.5 Theorem *If $V = A \oplus B$ and if p is the projection on A parallel to B then*
(1) $A = \operatorname{Im} p = \{x \in V \; ; \; x = p(x)\}$;
(2) $B = \operatorname{Ker} p$;
(3) *p is idempotent.*

Proof (1) It is clear that $A = \operatorname{Im} p \supseteq \{x \in V \; ; \; x = p(x)\}$. If now $a \in A$ then its unique representation as the sum of an element in A and an element in B is $a = a + 0$. Consequently $p(a) = a$ and the inclusion becomes equality.

(2) Let $x \in V$ have the unique representation $x = a + b$ where $a \in A$ and $b \in B$. Then since $p(x) = a$ we have
$$p(x) = 0_V \iff a = 0_V \iff x = b \in B.$$
In other words, $\operatorname{Ker} p = B$.

(3) For every $x \in V$ we have $f(x) \in A$ and so, by (1), $f(x) = f[f(x)]$. Thus $f = f \circ f$. $\diamond$

2.6 Theorem *A linear mapping $f : V \to V$ is a projection if and only if it is idempotent, in which case $V = \operatorname{Im} f \oplus \operatorname{Ker} f$ and f is the projection on $\operatorname{Im} f$ parallel to $\operatorname{Ker} f$.*

Proof Suppose that f is a projection. Then there exist subspaces A, B with $V = A \oplus B$ and f is the projection on A parallel to B. By 2.5, f is idempotent.

Conversely, suppose that $f : V \to V$ is idempotent. If $x \in \operatorname{Im} f \cap \operatorname{Ker} f$ then we have $x = f(y)$ for some y, and $f(x) = 0_V$. Consequently, $x = f(y) = f[f(y)] = f(x) = 0_V$ and hence
$$\operatorname{Im} f \cap \operatorname{Ker} f = \{0_V\}.$$
Now for every $x \in V$ we observe that
$$f[x - f(x)] = f(x) - f[f(x)] = f(x) - f(x) = 0_V$$
and so $x - f(x) \in \operatorname{Ker} f$. The identity $x = f(x) + x - f(x)$ now shows that
$$V = \operatorname{Im} f + \operatorname{Ker} f.$$
It follows by 2.2 that
$$V = \operatorname{Im} f \oplus \operatorname{Ker} f.$$
Suppose now that $x = a + b$ where $a \in \operatorname{Im} f$ and $b \in \operatorname{Ker} f$. Then $a = f(y)$ for some y, and $f(b) = 0_V$. Consequently,
$$f(x) = f(a+b) = f(a) + 0_V = f[f(y)] = f(y) = a.$$
In other words, f is the projection on $\operatorname{Im} f$ parallel to $\operatorname{Ker} f$. $\diamond$

2.7 Corollary *If $f : V \to V$ is a projection then so is $\mathrm{id}_V - f$. Moreover, in this case,*
$$\mathrm{Im}\, f = \mathrm{Ker}(\mathrm{id}_V - f).$$

Proof Writing $f \circ f = f^2$ we deduce from $f^2 = f$ that
$$(\mathrm{id}_V - f)^2 = \mathrm{id}_V - f - f + f^2 = \mathrm{id}_V - f.$$
Also, by 2.5, we have
$$x \in \mathrm{Im}\, f \iff x = f(x) \iff (\mathrm{id}_V - f)(x) = 0_V$$
and so $\mathrm{Im}\, f = \mathrm{Ker}(\mathrm{id}_V - f)$. $\diamondsuit$

We shall now show how the decomposition of a vector space into a direct sum of finitely many non-zero subspaces may be expressed in terms of projections.

2.8 Theorem *If V is a vector space then there are non-zero subspaces $V_1, \ldots, V_n$ of V such that $V = \bigoplus\limits_{i=1}^{n} V_i$ if and only if there are non-zero linear mappings $p_1, \ldots, p_n : V \to V$ such that*

(1) $\sum\limits_{i=1}^{n} p_i = \mathrm{id}_V$;

(2) $(i \neq j) \quad p_i \circ p_j = 0$.

Moreover, such mappings p_i are necessarily projections and $V_i = \mathrm{Im}\, p_i$ for $i = 1, \ldots, n$.

Proof Suppose first that $V = \bigoplus\limits_{i=1}^{n} V_i$. Then for $i = 1, \ldots, n$ we have $V = V_i \oplus \sum\limits_{j \neq i} V_j$. Let p_i be the projection on V_i parallel to $\sum\limits_{j \neq i} V_j$, and let $p_i^{\to}(X) = \{p_i(x)\ ;\ x \in X\}$ for every subspace X of V. Then for every $x \in V$ we have, for $j \neq i$,

$$\begin{aligned} p_i[p_j(x)] \in p_i^{\to}(\mathrm{Im}\, p_j) &= p_i^{\to}(V_j) \quad \text{by 2.6} \\ &\subseteq p_i^{\to}\left(\sum_{k \neq i} V_k\right) \\ &= p_i^{\to}(\mathrm{Ker}\, p_i) \quad \text{by 2.6} \\ &= \{0_V\} \end{aligned}$$

and so $p_i \circ p_j = 0$. Also, since every $x \in V$ can be written uniquely in the form $x = \sum_{i=1}^{n} x_i$ where $x_i \in V_i$ for each i, and since $p_i(x) = x_i$ for each i, we have

$$x = \sum_{i=1}^{n} x_i = \sum_{i=1}^{n} p_i(x) = \Big(\sum_{i=1}^{n} p_i\Big)(x)$$

whence $\sum_{i=1}^{n} p_i = \mathrm{id}_V$.

Conversely, suppose that $p_1, \ldots, p_n$ satisfy (1) and (2). Then we note that

$$p_i = p_i \circ \mathrm{id}_V = p_i \circ \sum_{j=1}^{n} p_j = \sum_{j=1}^{n} (p_i \circ p_j) = p_i \circ p_i$$

so each p_i is idempotent and therefore, by 2.6, is a projection. Now for every $x \in V$ we have

$$x = \mathrm{id}_V(x) = \Big(\sum_{i=1}^{n} p_i\Big)(x) = \sum_{i=1}^{n} p_i(x) \in \sum_{i=1}^{n} \mathrm{Im}\, p_i$$

which shows that $V = \sum_{i=1}^{n} \mathrm{Im}\, p_i$. If now $x \in \mathrm{Im}\, p_i \cap \sum_{j \neq i} \mathrm{Im}\, p_j$ then, by 2.5, $x = p_i(x)$ and $x = \sum_{j \neq i} x_j$ where $p_j(x_j) = x_j$ for every $j \neq i$. Consequently,

$$x = p_i(x) = p_i\Big(\sum_{j \neq i} x_j\Big) = p_i\Big(\sum_{j \neq i} p_j(x_j)\Big) = \sum_{j \neq i} p_i[p_j(x)] = 0_V$$

and it follows that $V = \bigoplus_{i=1}^{n} \mathrm{Im}\, p_i$. $\diamond$

The description in 2.8 opens the door to a deep study of linear mappings and their representation by matrices. In order to embark on this, we require the following notion.

Definition If V is a vector space over a field F and if $f : V \to V$ is linear then a subspace W of V is said to be *f-invariant* (or *f-stable*) if it satisfies the property

$$x \in W \implies f(x) \in W.$$

Example If $f : V \to V$ is linear then $\mathrm{Im}\, f$ and $\mathrm{Ker}\, f$ are f-invariant.

Example Let $D : \mathbb{R}[X] \to \mathbb{R}[X]$ be the differentiation map on the vector space of all real polynomials. Then the subspace $\mathbb{R}_n[X]$ of polynomials of degree at most $n+1$ is D-invariant.

Example If $f : V \to V$ is linear and $x \in V$ with $x \neq 0_V$ then the subspace spanned by $\{x\}$ is f-invariant if and only if x is an eigenvector of f. In fact, the subspace spanned by $\{x\}$ is $Fx = \{\lambda x \; ; \; \lambda \in F\}$ and this is f-invariant if and only if for every $\lambda \in F$ there exists $\mu \in F$ such that $f(\lambda x) = \mu x$. Taking $\lambda = 1_F$ we see that x is an eigenvector of f. Conversely, if x is an eigenvector of f then $f(x) = \mu x$ for some $\mu \in F$ and so, for every $\lambda \in F$, we have $f(\lambda x) = \lambda f(x) = \lambda \mu x$.

A useful result concerning invariant subspaces is the following.

2.9 Theorem *If $f : V \to V$ is linear then for every polynomial p over F the subspaces $\operatorname{Im} p(f)$ and $\operatorname{Ker} p(f)$ are f-invariant.*

Proof Observe that for every polynomial p we have

$$f \circ p(f) = p(f) \circ f.$$

It follows from this that if $x = p(f)(y)$ then $f(x) = p(f)[f(y)]$, so $\operatorname{Im} p(f)$ is f-invariant; and if $p(f)(x) = 0_V$ then $p(f)[f(x)] = 0_V$, so $\operatorname{Ker} p(f)$ is f-invariant. $\Diamond$

In what follows we shall often have occasion to deal with expressions of the form $p(f)$ where p is a polynomial and f is a linear mapping, and in so doing we shall find it convenient to denote composites by simple juxtaposition. Thus, for example, we shall write $fp(f)$ for $f \circ p(f)$, fg for $f \circ g$, f^2 for $f \circ f$.

Suppose now that V is of finite dimension n and that the subspace W of V is f-invariant. Choose a basis $\{w_1, \ldots, w_r\}$ of W and extend it to a basis

$$B = \{w_1, \ldots, w_r, v_1, \ldots, v_{n-r}\}$$

of V. Then, since W is f-invariant, it is readily seen that the matrix of f relative to B is of the form

$$\begin{bmatrix} A & B \\ 0 & C \end{bmatrix}$$

where A is an $r \times r$ matrix that represents the mapping induced on W by f.

Suppose now that $V = W_1 \oplus W_2$ where W_1 and W_2 are each f-invariant. If B_1 is a basis of W_1 and B_2 is a basis of W_2 then by 2.3 we have that $B = B_1 \cup B_2$ is a basis of V, and it is readily seen that the matrix of f relative to B is of the form

$$\begin{bmatrix} A_1 & 0 \\ 0 & A_2 \end{bmatrix}$$

where A_1, A_2 represent the mappings induced on W_1, W_2 by f.

More generally, if $V = \bigoplus_{i=1}^{n} W_i$ where each W_i is f-invariant and if B_i is a basis of W_i for each i then the matrix of f relative to the basis $B = \bigcup_{i=1}^{n} B_i$ is of the *block diagonal* form

$$\begin{bmatrix} A_1 & & & \\ & A_2 & & \\ & & \ddots & \\ & & & A_n \end{bmatrix}$$

in which A_i is the matrix representing the mapping induced on W_i by f, so that A_i is of size $\dim W_i \times \dim W_i$.

Our objective now is to use the notions of direct sum and invariant subspace in order to find a basis of V such that the matrix of f relative to this basis is as 'simple' (or as useful) as possible. The key to this study is the following result.

2.10 Theorem [Primary Decomposition] *Let V be a non-zero finite-dimensional vector space over a field F and let $f : V \to V$ be linear. Let the characteristic and minimum polynomials of f be*

$$\chi_f = p_1^{d_1} p_2^{d_2} \cdots p_k^{d_k}, \qquad m_f = p_1^{e_1} p_2^{e_2} \cdots p_k^{e_k}$$

respectively, where $p_1, \ldots, p_k$ are distinct irreducibles in $F[X]$. Then each of the subspaces $V_i = \operatorname{Ker} p_i(f)^{e_i}$ is f-invariant and $V = \bigoplus_{i=1}^{k} V_i$. Moreover, if $f_i : V_i \to V_i$ is the linear mapping induced on V_i by f then the minimum polynomial of f_i is $p_i^{e_i}$ and the characteristic polynomial of f_i is $p_i^{d_i}$.

Proof If $k = 1$ the result is trivial, so suppose that $k \geq 2$. For $i = 1, \ldots, k$ let $q_i = m_f/p_i^{e_i} = \prod_{j \neq i} p_j^{e_j}$. Then there is no irreducible factor that is common to each of $q_1, \ldots, q_k$ and so there exist $a_1, \ldots, a_n \in F[X]$ such that

$$q_1 a_1 + q_2 a_2 + \cdots + q_k a_k = 1.$$

Writing $t_i = q_i a_i$ for each i and substituting f in this polynomial identity, we obtain

(1) $\qquad t_1(f) + t_2(f) + \cdots + t_k(f) = \mathrm{id}_V .$

Now by the definition of q_i we have that if $i \neq j$ then m_f divides $q_i q_j$. Consequently $q_i(f) q_j(f) = 0$ for $i \neq j$ and then

(2) $\qquad (i \neq j) \qquad t_i(f) t_j(f) = 0.$

By (1), (2) and 2.8 we see that each $t_i(f)$ is a projection and

$$V = \bigoplus_{i=1}^k \mathrm{Im}\, t_i(f).$$

Moreover, by 2.9, each of the subspaces $\mathrm{Im}\, t_i(f)$ is f-invariant. We now show that $\mathrm{Im}\, t_i(f) = \mathrm{Ker}\, p_i(f)^{e_i}$.

Since $p_i^{e_i} q_i = m_f$ we have $p_i(f)^{e_i} q_i(f) = m_f(f) = 0$ from which it follows that $p_i(f)^{e_i} t_i(f) = 0$ and hence $\mathrm{Im}\, t_i(f) \subseteq \mathrm{Ker}\, p_i(f)^{e_i}$.

To establish the reverse inclusion, observe that, for every j,

$$t_j(f) = q_j(f) a_j(f) = \prod_{i \neq j} p_i(f)^{e_i} \cdot a_j(f)$$

and hence

$$\begin{aligned} \mathrm{Ker}\, p_i(f)^{e_i} &\subseteq \bigcap_{j \neq i} \mathrm{Ker}\, t_j(f) \\ &\subseteq \mathrm{Ker} \sum_{j \neq i} t_j(f) \\ &= \mathrm{Ker}(\mathrm{id}_V - t_i(f)) \qquad \text{by (1)} \\ &= \mathrm{Im}\, t_i(f) \qquad \text{by 2.7.} \end{aligned}$$

As for the induced mapping $f_i : V_i \to V_i$, let m_{f_i} be its minimum polynomial. Since $p_i(f)^{e_i}$ is the zero map on V_i, so is $p_i(f_i)^{e_i}$. Consequently we have that $m_{f_i}|p_i^{e_i}$. Thus $m_{f_i}|m_f$ and the m_{f_i} are relatively prime. Suppose now that $g \in F[X]$ is a multiple of m_{f_i} for every i. Then $g(f_i)$ is the zero map on V_i. If now $x = \sum_{i=1}^{k} v_i \in \bigoplus_{i=1}^{k} V_i = V$ then

$$g(f)(x) = \sum_{i=1}^{k} g(f)(v_i) = \sum_{i=1}^{k} g(f_i)(v_i) = 0_V$$

and so $g(f) = 0$ and consequently $m_f|g$. Thus we see that m_f is the least common multiple of $m_{f_1}, \ldots, m_{f_k}$. Since these polynomials are relatively prime, we then have $m_f = \prod_{i=1}^{k} m_{f_i}$. But we know that $m_f = \prod_{i=1}^{k} p_i^{e_i}$, and that $m_{f_i}|p_i^{e_i}$. Since all the polynomials in question are monic it follows that $m_{f_i} = p_i^{e_i}$ for $i = 1, \ldots, k$.

Finally, we can paste together bases of the subspaces V_i to form a basis of V with respect to which the matrix of f is of the block diagonal form

$$M = \begin{bmatrix} A_1 & & & \\ & A_2 & & \\ & & \ddots & \\ & & & A_k \end{bmatrix}.$$

Since, by the theory of determinants,

$$\det(XI - M) = \prod_{i=1}^{k} \det(XI - A_i)$$

we see that $\chi_f = \prod_{i=1}^{k} \chi_{f_i}$. Now we know that $m_{f_i} = p_i^{e_i}$ and so, by 1.4, we must have $\chi_f = p_i^{r_i}$ for some $r_i \geq e_i$. Thus

$$\prod_{i=1}^{k} p_i^{d_i} = \chi_f = \prod_{i=1}^{k} p_i^{r_i}$$

from which it follows that $r_i = d_i$ for $i = 1, \ldots, k$. $\diamond$

2.11 Corollary $(i = 1, \ldots, k)$ $\dim V_i = d_i \deg p_i$.

Proof $\dim V_i$ is the degree of χ_{f_i}. $\diamond$

2.12 Corollary *Let V be a non-zero finite-dimensional vector space over a field F. If $f : V \to V$ is linear and all the eigenvalues of f lie in F, so that*

$$\chi_f = (X - \lambda_1)^{d_1}(X - \lambda_2)^{d_2} \cdots (X - \lambda_k)^{d_k},$$
$$m_f = (X - \lambda_1)^{e_1}(X - \lambda_2)^{e_2} \cdots (X - \lambda_k)^{e_k},$$

then each of the subspaces $V_i = \operatorname{Ker}(f - \lambda_i \operatorname{id}_V)^{e_i}$ is f-invariant, of dimension d_i, and $V = \bigoplus_{i=1}^{k} V_i$. $\diamond$

Example Consider the linear mapping $f : \mathbb{R}^3 \to \mathbb{R}^3$ given by

$$f(x, y, z) = (-z, x + z, y + z).$$

Relative to the standard ordered basis, the matrix of f is

$$A = \begin{bmatrix} 0 & 0 & -1 \\ 1 & 0 & 1 \\ 0 & 1 & 1 \end{bmatrix}.$$

It is readily seen that $\chi_A = m_A = (X + 1)(X - 1)^2$. By 2.12,

$$\mathbb{R}^3 = \operatorname{Ker}(f + \operatorname{id}_V) \oplus \operatorname{Ker}(f - \operatorname{id}_V)^2$$

with $\operatorname{Ker}(f + \operatorname{id}_V)$ of dimension 1 and $\operatorname{Ker}(f - \operatorname{id}_V)^2$ of dimension 2. Now

$$(f + \operatorname{id}_V)(x, y, z) = (x - z, x + y + z, y + 2z)$$

so a basis for $\operatorname{Ker}(f + \operatorname{id}_V)$ is $\{(1, -2, 1)\}$. Also,

$$(f - \operatorname{id}_V)^2(x, y, z) = (x - y + z, -2x + 2y - 2z, x - y + z)$$

so a basis for $\operatorname{Ker}(f - \operatorname{id}_V)^2$ is $\{(0, 1, 1), (1, 1, 0)\}$. Thus a basis for $\mathbb{R}^3$ with respect to which the matrix of f is in block diagonal form is

$$B = \{(1, -2, 1), (0, 1, 1), (1, 1, 0)\}.$$

The transition matrix from B to the standard basis is

$$P = \begin{bmatrix} 1 & 0 & 1 \\ -2 & 1 & 1 \\ 1 & 1 & 0 \end{bmatrix}$$

and the block diagonal form of A is then

$$P^{-1}AP = \begin{bmatrix} -1 & & \\ & 2 & 1 \\ & -1 & 0 \end{bmatrix}.$$

Example Consider the differential equation

$$(D^n + a_{n-1}D^{n-1} + \cdots + a_1 D + a_0)f = 0$$

with constant (complex) coefficients. Let V be the solution space, i.e. the set of all infinitely differentiable functions satisfying the equation. If

$$m = X^n + a_{n-1}X^{n-1} + \cdots + a_1 X + a_0$$

then over $\mathbb{C}$ we have

$$m = (X - \alpha_1)^{e_1}(X - \alpha_2)^{e_2} \cdots (X - \alpha_k)^{e_k}.$$

Then $D : V \to V$ is linear and its minimum polynomial is m. By 2.12, V is the direct sum of the solution spaces V_i of the differential equations

$$(D - \alpha_i \operatorname{id})^{e_i} f = 0.$$

Now the solutions of $(D - \alpha \operatorname{id})^n f = 0$ can be determined using the fact that, by a simple inductive argument,

$$(D - \alpha \operatorname{id})^n f = e^{\alpha t} D^n (e^{-\alpha t} f).$$

Thus f is a solution if and only if $D^n(e^{-\alpha t} f) = 0$, which is the case if and only if $e^{-\alpha t} f$ is a polynomial of degree at most $n-1$. A basis for the solution space of $(D - \alpha \operatorname{id})^n f = 0$ is then

$$\{e^{\alpha t}, te^{\alpha t}, \ldots, t^{n-1} e^{\alpha t}\}.$$

It is natural to consider the particular case of the Primary Decomposition Theorem in which the irreducible factors p_i of m_f are all linear and each $e_i = 1$. This gives the following important result.

2.13 Theorem *Let V be a non-zero finite-dimensional vector space over a field F. Then a linear mapping $f : V \to V$ is diagonalizable if and only if its minimum polynomial m_f is a product of distinct linear factors.*

Proof Suppose that
$$m_f = (X - \lambda_1)(X - \lambda_2) \cdots (X - \lambda_k)$$
where $\lambda_1, \ldots, \lambda_k \in F$ are distinct. By 2.12, V is the direct sum of the f-invariant subspaces $V_i = \text{Ker}(f - \lambda_i \,\text{id}_V)$. For every $x \in V_i$ we have $(f - \lambda_i \,\text{id}_V)(x) = 0$, so that $f(x) = \lambda_i x$. Thus every non-zero element of V_i is an eigenvector associated with the eigenvalue λ_i. By 2.3, we can paste together bases for $V_1, \ldots, V_k$ to form a basis for V. Thus V has a basis consisting of eigenvectors of f, so f is diagonalizable.

Conversely, suppose that V has a basis consisting of eigenvectors of f. Let $\lambda_1, \ldots, \lambda_k$ be the distinct eigenvalues of f and consider the polynomial
$$p = (X - \lambda_1)(X - \lambda_2) \cdots (X - \lambda_k).$$
Clearly, $p(f)$ maps every basis vector to 0_V and consequently $p(f) = 0$. The minimum polynomial m_f therefore divides p, and must coincide with p since every eigenvalue is a zero of m_f. $\diamond$

Example Consider the linear mapping $f : \mathbb{R}^3 \to \mathbb{R}^3$ given by
$$f(x, y, z) = (7x - y - 2z, -x + 7y + 2z, -2x + 2y + 10z).$$
Relative to the standard ordered basis of $\mathbb{R}^3$, the matrix of f is
$$A = \begin{bmatrix} 7 & -1 & -2 \\ -1 & 7 & 2 \\ -2 & 2 & 10 \end{bmatrix}.$$
It is readily seen that $\chi_A = (X - 6)^2(X - 12)$ and that $m_A = (X - 6)(X - 12)$ so, by 2.13, f is diagonalizable.

An interesting result concerning diagonalizable mappings that will be useful later is the following.

DIRECT SUMS OF SUBSPACES

2.14 Theorem *Let V be a non-zero finite-dimensional vector space over a field F and let $f, g : V \to V$ be diagonalizable linear mappings. Then f and g are simultaneously diagonalizable (in the sense that there is a basis of V consisting of eigenvectors of both f and g) if and only if $f \circ g = g \circ f$.*

Proof $\Rightarrow$: Suppose that there is a basis $\{v_1, \ldots, v_n\}$ of V such that each v_i is an eigenvector of both f and g. If $f(v_i) = \lambda_i v_i$ and $g(v_i) = \mu_i v_i$ then

$$f[g(v_i)] = \lambda_i \mu_i v_i = \mu_i \lambda_i v_i = g[f(v_i)].$$

Since $f \circ g$ and $g \circ f$ thus agree on a basis, it follows that they are equal.

$\Leftarrow$: Suppose that $f \circ g = g \circ f$. Since f is diagonalizable its minimum polynomial is of the form

$$m_f = (X - \lambda_1)(X - \lambda_2) \cdots (X - \lambda_k)$$

where $\lambda_1, \ldots, \lambda_k$ are the distinct eigenvalues of f. By 2.12 we have $V = \bigoplus_{i=1}^{k} V_i$ where $V_i = \mathrm{Ker}(f - \lambda_i \, \mathrm{id}_V)$. Since, by hypothesis, $f \circ g = g \circ f$ we have, for $v_i \in V_i$,

$$f[g(v_i)] = g[f(v_i)] = g(\lambda_i v_i) = \lambda_i g(v_i)$$

and so $g(v_i) \in V_i$. Thus each V_i is g-invariant. Now let $g_i : V_i \to V_i$ be the linear mapping thus induced by g. Since g is diagonalizable so is each g_i, for the minimum polynomial of g_i divides that of g. We can therefore find a basis B_i of V_i consisting of eigenvectors of g_i. Since every eigenvector of g_i is an eigenvector of g and since every element of V_i is an eigenvector of f, it follows that $\bigcup_{i=1}^{k} B_i$ is a basis of V consisting of eigenvectors of both f and g. $\diamond$

2.15 Corollary *Let A, B be $n \times n$ matrices over a field F. If A and B are diagonalizable then they are simultaneously diagonalizable (i.e. there is an invertible matrix P such that $P^{-1}AP$ and $P^{-1}BP$ are diagonal) if and only if $AB = BA$.* $\diamond$

CHAPTER THREE

Reduction to triangular form

Despite the fact that, in general, $f : V \to V$ does not have a diagonal matrix representation, it is possible to 'simplify' the matrix representation of f in several ways. In this Chapter we shall describe the 'easiest' of these. We shall be concerned with those linear mappings f whose minimum polynomial (and hence also whose characteristic polynomial) factorises completely as a product of (not necessarily distinct) linear factors. Of course, this always happens when the ground field is $\mathbb{C}$, so the results we shall prove will be valid for all linear mappings on a finite-dimensional complex vector space. Specifically, we shall show that for such a mapping f there is an ordered basis of V with respect to which the matrix of f is triangular.

In order to see how to proceed, we observe first that by 2.12 we can write V as a direct sum of the f-invariant subspaces $V_i = \text{Ker}(f - \lambda_i \,\text{id}_V)^{e_i}$. Let $f_i : V_i \to V_i$ be the linear mapping induced on the 'primary component' V_i by f, and consider the mapping $f_i - \lambda_i \,\text{id}_{V_i} : V_i \to V_i$. We have that $(f_i - \lambda_i \,\text{id}_{V_i})^{e_i}$ is the zero map on V_i, so $f_i - \lambda_i \,\text{id}_{V_i}$ is *nilpotent*, in the following sense.

Definition A linear mapping $f : V \to V$ is said to be *nilpotent* if $f^m = 0$ for some positive integer m.

Example $f : \mathbb{R}^3 \to \mathbb{R}^3$ given by $f(x,y,z) = (0,x,y)$ is nilpotent. In fact, $f^2(x,y,z) = (0,0,x)$ and $f^3 = 0$.

Example If $f : \mathbb{C}^n \to \mathbb{C}^n$ is such that all the eigenvalues of f are 0 then $\chi_f = X^n$ and so, by Cayley-Hamilton, $f^n = \chi_f(f) = 0$. Thus f is nilpotent.

REDUCTION TO TRIANGULAR FORM

Example The differentiation map $D : \mathbb{R}_n[X] \to \mathbb{R}_n[X]$ is nilpotent.

We now produce a particularly useful basis for V in the presence of a nilpotent linear map.

3.1 Theorem *Let V be a non-zero finite-dimensional vector space over a field F and let $f : V \to V$ be a nilpotent linear mapping. Then there is a basis $\{v_1, \ldots, v_n\}$ of V such that*

$$f(v_1) = 0_V;$$
$$f(v_2) \in \langle v_1 \rangle;$$
$$f(v_3) \in \langle v_1, v_2 \rangle;$$
$$\vdots$$
$$f(v_n) \in \langle v_1, \ldots, v_{n-1} \rangle.$$

Proof Since f is nilpotent there is a positive integer m such that $f^m = 0$. If $f = 0$ then every basis of V satisfies the stated conditions, so let $f \neq 0$. Now let k be the smallest positive integer such that $f^k = 0$. Then $f^i \neq 0$ for $1 \leq i \leq k-1$. Since $f^{k-1} \neq 0$ there exists $v \in V$ such that $f^{k-1}(v) \neq 0_V$. Let $v_1 = f^{k-1}(v)$ and observe that that $f(v_1) = 0_V$. We now proceed recursively. Suppose that we have been able to find $v_1, \ldots, v_r$ satisfying the conditions, and let $W = \langle v_1, \ldots, v_r \rangle$. If $W = V$ there is nothing more to prove. If $W \neq V$ there are two possibilities : either $\text{Im}\, f \subseteq W$ or $\text{Im}\, f \not\subseteq W$. In the former case, let v_{r+1} be any element of $V \setminus W$. In the latter case, there is a positive integer j such that $\text{Im}\, f^j \not\subseteq W$ and $\text{Im}\, f^{j+1} \subseteq W$; for clearly we have the chain

$$\{0_V\} = \text{Im}\, f^k \subseteq \text{Im}\, f^{k-1} \subseteq \cdots \subseteq \text{Im}\, f^2 \subseteq \text{Im}\, f.$$

In this case we choose $v_{r+1} \in \text{Im}\, f^j$ with $v_{r+1} \notin W$. Each of these choices is such that $\{v_1, \ldots, v_{r+1}\}$ is linearly independent with $f(v_{r+1}) \in W$. $\diamond$

3.2 Corollary *If $f : V \to V$ is nilpotent then there is a basis of V with respect to which the matrix of f is upper triangular with all diagonal entries 0.*

Proof By 3.1 we can find a basis $\{v_1, \ldots, v_n\}$ of V such that

$$f(v_1) = 0_V;$$
$$f(v_2) = a_{12}v_1;$$
$$f(v_3) = a_{13}v_1 + a_{23}v_2;$$
$$\vdots$$
$$f(v_n) = a_{1n}v_1 + a_{2n}v_2 + \cdots + a_{n-1,n}v_{n-1},$$

which shows that the matrix of f is the upper triangular matrix

$$\begin{bmatrix} 0 & a_{12} & a_{13} & \ldots & a_{1n} \\ 0 & 0 & a_{23} & \ldots & a_{2n} \\ \vdots & \vdots & \vdots & & \vdots \\ 0 & 0 & 0 & \ldots & a_{n-1,n} \\ 0 & 0 & 0 & \ldots & 0 \end{bmatrix}$$

in which all the diagonal entries are 0. ◇

We can apply the above results to the nilpotent linear mapping $g_i = f_i - \lambda_i \,\mathrm{id}_{V_i}$ on the direct summand V_i of dimension d_i. Since $f_i = g_i + \lambda_i \,\mathrm{id}_{V_i}$ we have

$$\mathrm{Mat}\, f_i = \mathrm{Mat}\, g_i + \lambda_i \,\mathrm{Mat}\,\mathrm{id}_{V_i}$$
$$= \begin{bmatrix} \lambda_i & a_{12} & a_{13} & \ldots & a_{1d_i} \\ 0 & \lambda_i & a_{23} & \ldots & a_{2d_i} \\ \vdots & \vdots & \vdots & & \vdots \\ 0 & 0 & 0 & \ldots & a_{d_i-1,d_i} \\ 0 & 0 & 0 & \ldots & \lambda_i \end{bmatrix}.$$

Consequently, we have the following result.

3.3 Theorem [Triangular Form] *Let V be a non-zero finite-dimensional vector space over a field F and let $f : V \to V$ be a linear mapping whose characteristic and minimum polynomials are*

$$\chi_f = \prod_{i=1}^{k} (X - \lambda_i)^{d_i}, \qquad m_f = \prod_{i=1}^{k} (X - \lambda_i)^{e_i}$$

for distinct $\lambda_1, \ldots, \lambda_k \in F$ and $e_i \leq d_i$. Then there is an ordered basis of V with respect to which the matrix of f is the block diagonal matrix

$$\begin{bmatrix} A_1 & & & \\ & A_2 & & \\ & & \ddots & \\ & & & A_k \end{bmatrix}$$

in which A_i is a $d_i \times d_i$ upper triangular matrix of the form

$$\begin{bmatrix} \lambda_i & ? & \cdots & ? \\ 0 & \lambda_i & \cdots & ? \\ \vdots & \vdots & & \vdots \\ 0 & 0 & \cdots & \lambda_i \end{bmatrix}. \quad \diamondsuit$$

3.4 Corollary *Every square matrix over $\mathbb{C}$ is similar to an upper triangular matrix.* $\diamondsuit$

Example Consider the linear mapping $f : \mathbb{R}^3 \to \mathbb{R}^3$ given by

$$f(x, y, z) = (x + z, 2y + z, -x + 3z).$$

Relative to the standard ordered basis, the matrix of f is

$$A = \begin{bmatrix} 1 & 0 & 1 \\ 0 & 2 & 1 \\ -1 & 0 & 3 \end{bmatrix}.$$

It is readily seen that $\chi_A = m_A = (X - 2)^3$. By 2.12, therefore,

$$\mathbb{R}^3 = \operatorname{Ker}(f - 2\operatorname{id})^3.$$

We now find a basis for $\mathbb{R}^3$ in the style of 3.1. First, note that

$$(f - 2\operatorname{id})(x, y, z) = (-x + z, z, -x + z).$$

We therefore choose, for example,

$$v_1 = (0, 1, 0) \in \operatorname{Ker}(f - 2\operatorname{id}) \setminus \{0\}.$$

As for v_2, we require v_2 independent of v_1 with $(f - 2\,\mathrm{id})(v_2) \in \langle v_1 \rangle$, i.e. we have to find $v_2 = (x, y, z)$ independent of v_1 such that
$$(-x + z, z, -x + z) = \alpha(0, 1, 0).$$
We can take, for example, $v_2 = (1, 0, 1)$. Now we require v_3 independent of v_1, v_2 and such that $(f - 2\,\mathrm{id})(v_3) \in \langle v_1, v_2 \rangle$. We can take, for example, $v_3 = (0, 0, 1)$. Consider now the basis
$$B = \{(0, 1, 0), (1, 0, 1), (0, 0, 1)\}.$$
The transition matrix from B to the standard basis is
$$P = \begin{bmatrix} 0 & 1 & 0 \\ 1 & 0 & 0 \\ 0 & 1 & 1 \end{bmatrix}$$
and so the matrix of f relative to B is the upper triangular matrix
$$P^{-1}AP = \begin{bmatrix} 2 & 1 & 1 \\ 0 & 2 & 1 \\ 0 & 0 & 2 \end{bmatrix}.$$

Example Consider the linear mapping $f : \mathbb{R}^3 \to \mathbb{R}^3$ given by
$$f(x, y, z) = (-z, x + z, y + z).$$
Relative to the standard ordered basis, the matrix of f is
$$A = \begin{bmatrix} 0 & 0 & -1 \\ 1 & 0 & 1 \\ 0 & 1 & 1 \end{bmatrix}.$$
It is readily seen that $\chi_A = m_A = (X + 1)(X - 1)^2$. By 2.12, therefore,
$$\mathbb{R}^3 = \mathrm{Ker}(f + \mathrm{id}) \oplus \mathrm{Ker}(f - \mathrm{id})^2,$$
where $\mathrm{Ker}(f + \mathrm{id})$ is of dimension 1 and $\mathrm{Ker}(f - \mathrm{id})^2$ is of dimension 2. Since
$$(f + \mathrm{id})(x, y, z) = (x - z, x + y + z, y + 2z),$$

a basis for $V_1 = \mathrm{Ker}(f+\mathrm{id})$ is $\{(1,-2,1)\}$. Now consider finding a basis for $V_2 = \mathrm{Ker}(f-\mathrm{id})^2$ in the style of 3.1. First note that
$$(f-\mathrm{id})(x,y,z) = (-x-z, x-y+z, y).$$
Begin by choosing
$$v_1 = (-1,0,1) \in \mathrm{Ker}(f-\mathrm{id}) \setminus \{0\}.$$
We now require v_2 independent of v_1 with $(f-\mathrm{id})(v_2) \in \langle v_1 \rangle$, i.e. we have to find $v_2 = (x,y,z)$ independent of v_1 such that
$$(-x-z, x-y+z, y) = \alpha(-1,0,1).$$
We can choose, for example, $v_2 = (0,1,1)$. Since
$$(f-\mathrm{id})(v_1) = (0,0,0) = 0v_1 + 0v_2$$
$$(f-\mathrm{id})(v_2) = (-1,0,1) = 1v_1 + 0v_2$$
we see that the matrix of $f - \mathrm{id}$ relative to the basis $\{v_1, v_2\}$ is
$$\begin{bmatrix} 0 & 1 \\ 0 & 0 \end{bmatrix}.$$
The matrix of the mapping f_2 induced on $V_2 = \mathrm{Ker}(f-\mathrm{id})^2$ by f is then
$$\begin{bmatrix} 1 & 1 \\ 0 & 1 \end{bmatrix}.$$
Consequently, the matrix of f relative to the basis
$$B = \{(1,-2,1), (-1,0,1), (0,1,1)\}$$
is the upper triangular matrix
$$\begin{bmatrix} -1 & 0 & 0 \\ 0 & 1 & 1 \\ 0 & 0 & 1 \end{bmatrix}.$$

We have seen above that each induced mapping f_i on the f-invariant subspace $V_i = \mathrm{Ker}(f - \lambda_i \,\mathrm{id}_V)^{e_i}$ can be written as $f_i = g_i + \lambda_i \,\mathrm{id}_{V_i}$ where g_i is nilpotent. Clearly, $\lambda_i \,\mathrm{id}_{V_i}$ is a diagonalizable mapping (its minimum polynomial being $X - \lambda_i$). So every induced mapping f_i has a decomposition in the form of a diagonalizable mapping plus a nilpotent mapping. This is true 'globally' of f itself, as we shall now show.

3.5 Theorem [Jordan Decomposition] *Let V be a non-zero finite-dimensional vector space over a field F and let $f : V \to V$ be a linear mapping all of whose eigenvalues lie in F. Then there is a diagonalizable linear mapping $\delta : V \to V$ and a nilpotent linear mapping $\eta : V \to V$ such that $f = \delta + \eta$ and $\delta \circ \eta = \eta \circ \delta$. Moreover, there are polynomials $p, q \in F[X]$ such that $\delta = p(f)$ and $\eta = q(f)$. Furthermore, δ and η are uniquely determined, in the sense that if $\delta', \eta' : V \to V$ are respectively diagonalizable and nilpotent linear mappings such that $f = \delta' + \eta'$ with $\delta' \circ \eta' = \eta' \circ \delta'$ then $\delta = \delta'$ and $\eta = \eta'$.*

Proof The minimum polynomial of f is
$$m_f = (X - \lambda_1)^{e_1}(X - \lambda_2)^{e_2} \cdots (X - \lambda_k)^{e_k}$$
where $\lambda_1, \ldots, \lambda_k \in F$ are distinct. By 2.12, we have $V = \bigoplus_{i=1}^{k} V_i$ where $V_i = \text{Ker}(f - \lambda_i \,\text{id}_V)^{e_i}$. Let $\delta : V \to V$ be given by $\delta = \sum_{i=1}^{k} \lambda_i p_i$ where $p_i : V \to V$ is the projection on V_i parallel to $\sum_{j \neq i} V_j$. Since, for $v_i \in V_i$, $\delta(v_i) = \left(\sum_{j=1}^{k} \lambda_j p_j\right)(v_i) = \lambda_i v_i$ it follows that V has a basis consisting of eigenvectors of δ and so δ is diagonalizable.

Now let $\eta = f - \delta$. Then for $v_i \in V_i$ we have
$$\eta(v_i) = f(v_i) - \delta(v_i) = (f - \lambda_i \,\text{id}_V)(v_i)$$
and consequently $\eta^{e_i}(v_i) = (f - \lambda_i \,\text{id}_V)^{e_i}(v_i) = 0_V$. It follows that, for some r, $\text{Ker}\,\eta^r$ contains a basis of V, so $\eta^r = 0$ and hence η is nilpotent.

Since $V = \bigoplus_{i=1}^{k} V_i$, every $v \in V$ can be written uniquely as $v = v_1 + \cdots + v_k$ with $v_i \in V_i$ for every i. Since each V_i is f-invariant,
$$p_i[f(v)] = p_i[f(v_1) + \cdots + f(v_k)] = f(v_i) = f[p_i(v)]$$
and hence $p_i \circ f = f \circ p_i$ for each i. Consequently
$$\delta \circ f = \left(\sum_{i=1}^{k} \lambda_i p_i\right) \circ f = \sum_{i=1}^{k} \lambda_i (p_i \circ f)$$
$$= \sum_{i=1}^{k} \lambda_i (f \circ p_i) = f \circ \left(\sum_{i=1}^{k} \lambda_i p_i\right) = f \circ \delta.$$

It follows from this that

$$\delta \circ \eta = \delta(f - \delta) = \delta f - \delta^2 = f\delta - \delta^2 = (f - \delta)\delta = \eta \circ \delta.$$

We now show that there are polynomials $p, q \in F[X]$ such that $\delta = p(f)$ and $\eta = q(f)$. Now it is clear that $p_i = t_i(f)$ where t_i is the polynomial described in the proof of 2.10. Thus, by the definition of δ, we have $\delta = p(f)$ where $p = \sum_{i=1}^{k} \lambda_i t_i$. Since $\eta = f - \delta$, there is then a polynomial q such that $\eta = q(f)$.

As for uniqueness, suppose that $\delta', \eta' : V \to V$ are diagonalizable and nilpotent respectively, with $f = \delta' + \eta'$ and $\delta' \circ \eta' = \eta' \circ \delta'$. Then we have $f \circ \delta' = \delta' \circ f$ and $f \circ \eta' = \eta' \circ f$. Now we have just seen that there are polynomials p, q such that $\delta = p(f)$ and $\eta = q(f)$. We deduce, therefore, that $\delta \circ \delta' = \delta' \circ \delta$ and that $\eta \circ \eta' = \eta' \circ \eta$. Now since $\delta + \eta = f = \delta' + \eta'$ we have the equality $\delta - \delta' = \eta' - \eta$ and so, since η, η' commute, we can use the binomial theorem to deduce from the fact that η and η' are nilpotent that $\eta' - \eta$ is also nilpotent and therefore can be represented by a nilpotent matrix N. Also, since δ and δ' commute, it follows by 2.14 that there is a basis of V consisting of eigenvectors of both δ and δ'. Each such eigenvector is then an eigenvector of $\delta - \delta'$ and consequently $\delta - \delta'$ is represented by a diagonal matrix D. Now N and D are similar, and the only possibility is clearly $N = D = 0$. Consequently we have $\delta - \delta' = 0 = \eta' - \eta$ as required. $\diamondsuit$

There is, of course, a corresponding result that can be stated in terms of square matrices.

Example In the previous Example we saw that if $f : \mathbb{R}^3 \to \mathbb{R}^3$ is given by

$$f(x, y, z) = (-z, x + z, y + z)$$

then relative to the basis $\{(1, -2, 1), (-1, 0, 1), (0, 1, 1)\}$ the matrix of f is

$$T = \begin{bmatrix} -1 & 0 & 0 \\ 0 & 1 & 1 \\ 0 & 0 & 1 \end{bmatrix}.$$

Clearly, we can write $T = D + N$ where

$$D = \begin{bmatrix} -1 & 0 & 0 \\ 0 & 1 & 0 \\ 0 & 0 & 1 \end{bmatrix}, \qquad N = \begin{bmatrix} 0 & 0 & 0 \\ 0 & 0 & 1 \\ 0 & 0 & 0 \end{bmatrix}$$

and this is the Jordan decomposition of T. To refer matters back to the standard ordered basis, we compute PDP^{-1} and PNP^{-1} where P is the transition matrix

$$P = \begin{bmatrix} 1 & -1 & 0 \\ -2 & 0 & 1 \\ 1 & 1 & 1 \end{bmatrix}.$$

It is readily seen that

$$PDP^{-1} = \begin{bmatrix} \frac{1}{2} & \frac{1}{2} & -\frac{1}{2} \\ 1 & 0 & 1 \\ -\frac{1}{2} & \frac{1}{2} & \frac{1}{2} \end{bmatrix}, \qquad PNP^{-1} = \begin{bmatrix} -\frac{1}{2} & -\frac{1}{2} & -\frac{1}{2} \\ 0 & 0 & 0 \\ \frac{1}{2} & \frac{1}{2} & \frac{1}{2} \end{bmatrix}.$$

Consequently, the diagonal part of f is given by

$$d_f(x,y,z) = (\tfrac{1}{2}x + \tfrac{1}{2}y - \tfrac{1}{2}z, x + z, -\tfrac{1}{2}x + \tfrac{1}{2}y + \tfrac{1}{2}z),$$

and the nilpotent part of f is given by

$$n_f(x,y,z) = (-\tfrac{1}{2}x - \tfrac{1}{2}y - \tfrac{1}{2}z, 0, \tfrac{1}{2}x + \tfrac{1}{2}y + \tfrac{1}{2}z).$$

The Jordan decomposition of f (or of A) is particularly useful in computing powers of f (or of A). Indeed, since $f = \delta + \eta$ where δ, η commute we can apply the binomial theorem to obtain

$$f^n = (\delta + \eta)^n = \sum_{i=0}^{n} \binom{n}{i} \delta^{n-i} \eta^i.$$

The powers of δ are easily computed (by considering the powers of the corresponding diagonal matrix), and all the powers of η from some point on are zero.

CHAPTER FOUR

Reduction to Jordan form

It is natural to ask whether we can improve on the triangular form. In order to do so, it is clearly necessary to find 'better' bases for the subspaces V_i that appear as the direct summands or 'primary components' in the Primary Decomposition Theorem. So let us take a closer look at nilpotent mappings. For this purpose, we note the following result.

4.1 Theorem *If $f : V \to V$ is linear then, for every positive integer i,*

(1) $\operatorname{Ker} f^i \subseteq \operatorname{Ker} f^{i+1}$;
(2) $x \in \operatorname{Ker} f^{i+1} \implies f(x) \in \operatorname{Ker} f^i$.

Proof (1) Clearly, if $f^i(x) = 0_V$ then $f^{i+1}(x) = f[f^i(x)] = 0_V$.
(2) If $f^{i+1}(x) = 0_V$ then $f^i[f(x)] = 0_V$. $\diamond$

Definition If $f : V \to V$ is nilpotent then the *index* of f is the least positive integer k such that $f^k = 0$.

4.2 Theorem *Let V be a non-zero vector space over a field F and let $f : V \to V$ be a linear mapping that is nilpotent of index k. Then there is the chain of distinct subspaces*

$$\{0_V\} \subset \operatorname{Ker} f \subset \operatorname{Ker} f^2 \subset \cdots \subset \operatorname{Ker} f^{k-1} \subset \operatorname{Ker} f^k = V.$$

Proof Since $f^k = 0$ it is clear that f is not invertible and so $\operatorname{Ker} f \neq \{0_V\}$. In view of 4.1 it suffices to prove that $\operatorname{Ker} f^i \neq \operatorname{Ker} f^{i+1}$ for $i = 1, \ldots, k-1$. Suppose that for some i we have $\operatorname{Ker} f^i = \operatorname{Ker} f^{i+1}$. Then for every $x \in V$ we have

$$0_V = f^k(x) = f^{i+1}[f^{k-(i+1)}(x)]$$

whence $f^{k-(i+1)}(x) \in \operatorname{Ker} f^{i+1} = \operatorname{Ker} f^i$ and so

$$0_V = f^i[f^{k-(i+1)}(x)] = f^{k-1}(x).$$

Thus we have the contradiction $f^{k-1} = 0$. The subspaces in the above chain are therefore distinct. $\diamond$

Concerning the chain in 4.2, we have the following simple, but important, observation.

4.3 Theorem *If, respectively,*

$$\{u_1, \ldots, u_r\},$$
$$\{u_1, \ldots, u_r, v_1, \ldots, v_s\},$$
$$\{u_1, \ldots, u_r, v_1, \ldots, v_s, w_1, \ldots, w_t\}$$

are bases for $\operatorname{Ker} f^{i-1}, \operatorname{Ker} f^i, \operatorname{Ker} f^{i+1}$ *then*

$$S = \{u_1, \ldots, u_r, f(w_1), \ldots, f(w_t)\}$$

is a linearly independent subset of $\operatorname{Ker} f^i$.

Proof If $x \in \operatorname{Ker} f^{i+1}$ then $0_V = f^{i+1}(x) = f^i[f(x)]$ gives $f(x) \in \operatorname{Ker} f^i$. Consequently we see that $S \subseteq \operatorname{Ker} f^i$. Suppose now that

$$(\star) \quad \lambda_1 u_1 + \cdots + \lambda_r u_r + \mu_1 f(w_1) + \cdots + \mu_t f(w_t) = 0_V.$$

Then we have

$$\sum_{j=1}^{t} \mu_j f(w_j) = -\sum_{k=1}^{r} \lambda_k u_k \in \operatorname{Ker} f^{i-1}$$

from which it follows that

$$\sum_{j=1}^{t} \mu_j w_j \in \operatorname{Ker} f^i.$$

The only possibility is $\sum_{j=1}^{t} \mu_j w_j = 0_V$ (since otherwise we would have a dependence relation between the above basis elements of $\operatorname{Ker} f^{i+1}$), and consequently each $\mu_j = 0$. Then, by $(\star)$, we obtain $\sum_{k=1}^{r} \lambda_k u_k = 0_V$ and hence also each $\lambda_k = 0$. Thus S is linearly independent. $\diamond$

Definition By an *elementary Jordan matrix* associated with $\lambda \in F$ we mean either the 1×1 matrix $[\lambda]$ or a square matrix of the form

$$\begin{bmatrix} \lambda & 1 & 0 & 0 & \cdots & 0 \\ 0 & \lambda & 1 & 0 & \cdots & 0 \\ 0 & 0 & \lambda & 1 & \cdots & 0 \\ \vdots & \vdots & \vdots & \vdots & \ddots & \vdots \\ 0 & 0 & 0 & 0 & \cdots & \lambda \end{bmatrix}$$

in which the diagonal entries are all λ, the entries immediately above diagonal elements are all 1, and all other entries are 0. By a *Jordan block* matrix associated with $\lambda \in F$ we mean a matrix of the form

$$\begin{bmatrix} J_1 & & & \\ & J_2 & & \\ & & \ddots & \\ & & & J_t \end{bmatrix}$$

where each J_i is an elementary Jordan matrix associated with λ.

4.4 Theorem *Let V be a non-zero finite-dimensional vector space over a field F and let $f : V \to V$ be nilpotent of index k. Then there is a basis of V with respect to which the matrix of f is a Jordan block.*

Proof For $i = 1, \ldots, k$ let $W_i = \operatorname{Ker} f^i$. Then by 4.2 we have the chain

$$\{0_V\} \subset W_1 \subset W_2 \subset \cdots \subset W_{k-1} \subset W_k = V.$$

Construct a basis of V as follows. Choose a basis B_1 of W_1 and then, by successive extensions, form a basis B_i for each W_i :

$$B_1 = \{b_1, \ldots, b_{n_1}\},$$
$$B_2 = \{b_1, \ldots, b_{n_1}, b_{n_1+1}, \ldots, b_{n_2}\},$$
$$\vdots$$
$$B_{k-1} = \{b_1, \ldots, b_{n_1}, \ldots, b_{n_{k-2}+1}, \ldots, b_{n_{k-1}}\},$$
$$B_k = \{b_1, \ldots, b_{n_1}, \ldots, b_{n_{k-2}+1}, \ldots, b_{n_{k-1}}, b_{n_{k-1}+1}, \ldots, b_{n_k}\}.$$

Now let
$$T_k = \{b_{n_{k-1}+1}, \ldots, b_{n_k}\}$$
and write $f^{\rightarrow}(T_k) = \{f(x) \; ; \; x \in T_k\}$. Then by 4.3 the set
$$B_{k-2} \cup f^{\rightarrow}(T_k)$$
is a linearly independent subset of W_{k-1}. Extend this to a basis
$$B_{k-2} \cup f^{\rightarrow}(T_k) \cup \{y_1, \ldots, y_\beta\}$$
of W_{k-1}. Now let
$$T_{k-1} = f^{\rightarrow}(T_k) \cup \{y_1, \ldots, y_\beta\}.$$
Then by 4.3 the set
$$B_{k-3} \cup f^{\rightarrow}(T_{k-1})$$
is a linearly independent subset of W_{k-2}. Extend this to a basis
$$B_{k-3} \cup f^{\rightarrow}(T_{k-1}) \cup \{z_1, \ldots, z_\gamma\}$$
of W_{k-2}, and so on. Writing T_k as $\{x_1, \ldots, x_\alpha\}$, we thus see that we can form the following basis of V :

$$\begin{array}{l} x_1, \quad \ldots, \quad x_\alpha, \\ f(x_1), \quad \ldots, \quad f(x_\alpha), \quad y_1, \quad \ldots, \quad y_\beta, \\ f^2(x_1), \quad \ldots, \quad f^2(x_\alpha), \quad f(y_1), \quad \ldots, \quad f(y_\beta), \quad z_1, \ldots, z_\gamma, \\ \vdots \\ f^{k-1}(x_1), \ldots, f^{k-1}(x_\alpha), f^{k-2}(y_1), \ldots, f^{k-2}(y_\beta), \ldots, q_1, \ldots, q_\omega. \end{array}$$

Note that in this table the elements in the i-th row from the bottom are in W_i. Also, every element in the table is mapped by f to the element lying immediately below it, the elements in the bottom row being mapped to 0_V. Now order this basis by taking the first column starting at the bottom, then the second column starting at the bottom, and so on. Then it is readily seen that the ordered basis B that we obtain in this way is such that the matrix of f relative to B is a Jordan block associated with the eigenvalue 0. $\diamond$

REDUCTION TO JORDAN FORM

Example To illustrate the above argument, consider the mapping $f : \mathbb{R}^4 \to \mathbb{R}^4$ given by

$$f(a,b,c,d) = (0,a,d,0).$$

We have $f^2 = 0$ so f is nilpotent of index 2. Now

$$V_1 = \operatorname{Ker} f = \{(0,b,c,0) \; ; \; b,c \in \mathbb{R}\}, \qquad V_2 = \operatorname{Ker} f^2 = \mathbb{R}^4.$$

A basis for V_1 is $B_1 = \{(0,1,0,0),(0,0,1,0)\}$ which we extend to a basis

$$B_2 = \{(0,1,0,0),(0,0,1,0),(1,0,0,0),(0,0,0,1)\}$$

of $\mathbb{R}^4$. Now consider $T_2 = \{(1,0,0,0),(0,0,0,1)\}$. We have

$$f^{\to}(T_2) = \{(0,1,0,0),(0,0,1,0)\}$$

and $B_1 \cup f^{\to}(T_2) = B_1$. We then form the basis

$$(1,0,0,0), (0,0,0,1),$$
$$(0,1,0,0), (0,0,1,0)$$

of $\mathbb{R}^4$ and order it as follows :

$$B = \{(0,1,0,0),(1,0,0,0),(0,0,1,0),(0,0,0,1)\}.$$

The transition matrix from B to the standard basis is

$$P = \begin{bmatrix} 0 & 1 & 0 & 0 \\ 1 & 0 & 0 & 0 \\ 0 & 0 & 1 & 0 \\ 0 & 0 & 0 & 1 \end{bmatrix}.$$

Now $P^{-1} = P$ and the matrix of f relative to the standard basis is

$$A = \begin{bmatrix} 0 & 0 & 0 & 0 \\ 1 & 0 & 0 & 0 \\ 0 & 0 & 0 & 1 \\ 0 & 0 & 0 & 0 \end{bmatrix}.$$

So the Jordan block is given by

$$P^{-1}AP = \begin{bmatrix} 0 & 1 & & \\ 0 & 0 & & \\ & & 0 & 1 \\ & & 0 & 0 \end{bmatrix}.$$

In practice, we rarely have to carry out the above computation. To discover why, let us look more closely at the proof of 4.4. Observe that

$$|T_k| = \alpha = n_k - n_{k-1},$$
$$|T_{k-1}| = \alpha + \beta = n_{k-1} - n_{k-2},$$
$$|T_{k-2}| = \alpha + \beta + \gamma = n_{k-2} - n_{k-3},$$
$$\vdots$$

and consequently

$$\alpha + \beta + \gamma + \cdots + \omega = \dim \operatorname{Ker} f,$$

as can be seen by referring to the basis displayed on page 34. The number of elements in the bottom row of this display is $\dim \operatorname{Ker} f$. Now from this basis it is clear that there are $\alpha \geq 1$ elementary Jordan matrices of size $k \times k$ involved, then $\beta \geq 0$ of size $(k-1) \times (k-1)$, and so on. So we conclude from the above observation that *the number of elementary Jordan matrices appearing is* $\dim \operatorname{Ker} f$.

Returning to our Example, we see that $\operatorname{Ker} f$ has dimension 2, so there are precisely two elementary Jordan matrices involved. Since one at least has to be of size $k \times k = 2 \times 2$, the only possibility for the Jordan block is

$$\begin{bmatrix} 0 & 1 & & \\ 0 & 0 & & \\ & & 0 & 1 \\ & & 0 & 0 \end{bmatrix}.$$

Let us now apply 4.4 to the mappings $f_i - \lambda_i \operatorname{id}_{V_i}$ of 2.12. Note that by 2.10 the minimum polynomial of f_i is $m_{f_i} = (X - \lambda_i)^{e_i}$. Consequently we have that *the mapping $f_i - \lambda_i \operatorname{id}_{V_i}$ is nilpotent of index e_i on the d_i-dimensional subspace V_i*.

4.5 Theorem *Let V be a non-zero finite-dimensional vector space over a field F. If $f : V \to V$ is linear and if all the eigenvalues of f lie in F then there is an ordered basis of V with respect to which the matrix of f is a block diagonal matrix*

$$\begin{bmatrix} A_1 & & & \\ & A_2 & & \\ & & \ddots & \\ & & & A_k \end{bmatrix}$$

in which every A_i is a Jordan block.

Proof With the usual notation, if we apply 4.4 to the nilpotent mapping $f_i - \lambda_i \operatorname{id}_{V_i}$ then we see that there is a basis of $V_i = \operatorname{Ker}(f - \lambda_i \operatorname{id}_V)^{e_i}$ with respect to which the matrix of $f_i - \lambda_i \operatorname{id}_{V_i}$ is a Jordan block with 0 down the diagonal (since the only eigenvalue of a nilpotent mapping is 0). It follows that the matrix of f_i is a Jordan block with λ_i down the diagonal. ◇

Definition A matrix of the form described in 4.5 is called a *Jordan canonical matrix* of f.

Of course a Jordan canonical matrix is, strictly speaking, not unique since the order in which the Jordan blocks A_i appear down the diagonal is not specified. However, the number of such blocks, the size of each block, and the number of elementary Jordan matrices that appear in each block, are uniquely determined by f. So, provided we ignore the order of the blocks, we can choose to talk of 'the' Jordan matrix that represents f. This is often also called the *Jordan normal form*.

If the characteristic and minimum polynomials of f are

$$\chi_f = \prod_{i=1}^{k} (X - \lambda_i)^{d_i}, \qquad m_f = \prod_{i=1}^{k} (X - \lambda_i)^{e_i}$$

then from the previous discussion we have that, in the Jordan form, the eigenvalue λ_i appears d_i times in the diagonal, and the number of elementary Jordan matrices associated with λ_i is $\dim \operatorname{Ker}(f_i - \lambda_i \operatorname{id}_{V_i})$, which is the geometric multiplicity of the eigenvalue λ_i. Moreover, at least one of these elementary Jordan matrices is of size $e_i \times e_i$.

Example Let $f : \mathbb{R}^7 \to \mathbb{R}^7$ be linear with characteristic and minimum polynomials

$$\chi_f = (X-1)^3(X-2)^4, \qquad m_f = (X-1)^2(X-2)^3.$$

In any Jordan matrix that represents f the eigenvalue 1 appears three times in the diagonal, with at least one associated elementary Jordan matrix being of size 2×2; and the eigenvalue 2 appears four times in the diagonal, with at least one associated elementary Jordan matrix of size 3×3. Up to the order of the blocks, the only possibility is therefore

$$\begin{bmatrix} 1 & 1 & & & & & \\ & 1 & & & & & \\ & & 1 & & & & \\ & & & 2 & 1 & & \\ & & & & 2 & 1 & \\ & & & & & 2 & \\ & & & & & & 2 \end{bmatrix}.$$

Example Let us modify the previous Example slightly. Suppose that χ_f is as before and that now

$$m_f = (X-1)^2(X-2)^2.$$

In this case the eigenvalue 2 appears four times in the diagonal with at least one associated elementary Jordan matrix of size 2×2. The possibilities for the Jordan form are then

$$\begin{bmatrix} 1 & 1 & & & & & \\ & 1 & & & & & \\ & & 1 & & & & \\ & & & 2 & 1 & & \\ & & & & 2 & & \\ & & & & & 2 & 1 \\ & & & & & & 2 \end{bmatrix}, \begin{bmatrix} 1 & 1 & & & & & \\ & 1 & & & & & \\ & & 1 & & & & \\ & & & 2 & 1 & & \\ & & & & 2 & & \\ & & & & & 2 & \\ & & & & & & 2 \end{bmatrix}.$$

Example If $f : V \to V$ has characteristic polynomial

$$\chi_f = (X-2)^2(X-3)^3$$

then the possible Jordan forms, obtained by considering all six possible minimum polynomials, are

$$\begin{bmatrix} A & \\ & B \end{bmatrix}$$

where A is one of

$$\begin{bmatrix} 2 & 1 \\ & 2 \end{bmatrix}, \quad \begin{bmatrix} 2 & \\ & 2 \end{bmatrix}$$

and B is one of

$$\begin{bmatrix} 3 & 1 & \\ & 3 & 1 \\ & & 3 \end{bmatrix}, \quad \begin{bmatrix} 3 & 1 & \\ & 3 & \\ & & 3 \end{bmatrix}, \quad \begin{bmatrix} 3 & & \\ & 3 & \\ & & 3 \end{bmatrix}.$$

We now consider the problem of finding a *Jordan basis* for f, i.e. a basis of V with respect to which the matrix of f is a Jordan canonical matrix J. This is, of course, equivalent to the problem of finding an invertible matrix P such that $P^{-1}AP = J$ where A is the matrix that represents f relative to some fixed ordered basis. To see how to proceed, it suffices to consider the very special case where the Jordan matrix of f is the $t \times t$ matrix

$$\begin{bmatrix} \lambda & 1 & & & & \\ & \lambda & 1 & & & \\ & & \lambda & 1 & & \\ & & & \ddots & \ddots & \\ & & & & \lambda & 1 \\ & & & & & \lambda \end{bmatrix}.$$

A corresponding basis $\{v_1, \ldots, v_t\}$ will be such that

$$f(v_1) = \lambda v_1,$$
$$f(v_2) = \lambda v_2 + v_1,$$
$$f(v_3) = \lambda v_3 + v_2,$$
$$\vdots$$
$$f(v_{t-1}) = \lambda v_{t-1} + v_{t-2},$$
$$f(v_t) = \lambda v_t + v_{t-1}.$$

Thus, for every $t \times t$ elementary Jordan matrix associated with λ we require $v_1, \ldots, v_t$ to be linearly independent with

$$v_1 \in \operatorname{Im}(f - \lambda \operatorname{id}) \cap \operatorname{Ker}(f - \lambda \operatorname{id});$$
$$(i = 2, \ldots, t) \quad (f - \lambda \operatorname{id})(v_i) = v_{i-1}.$$

Example Let $f : \mathbb{R}^3 \to \mathbb{R}^3$ be given by

$$f(x, y, z) = (x + y, -x + 3y, -x + y + 2z).$$

Relative to the standard ordered basis, the matrix of f is

$$A = \begin{bmatrix} 1 & 1 & 0 \\ -1 & 3 & 0 \\ -1 & 1 & 2 \end{bmatrix}.$$

We have $\chi_A = (X - 2)^3$ and $m_A = (X - 2)^2$. The Jordan form is then

$$J = \begin{bmatrix} 2 & 1 & \\ & 2 & \\ & & 2 \end{bmatrix}.$$

Now we have

$$(f - 2 \operatorname{id})(x, y, z) = (-x + y, -x + y, -x + y)$$

and we have to first choose $v_1 \in \operatorname{Im}(f - 2 \operatorname{id}) \cap \operatorname{Ker}(f - 2 \operatorname{id})$. Clearly, $v_1 = (1, 1, 1)$ will do. Next we have to find v_2, independent of v_1, such that $(f - 2 \operatorname{id})(v_2) = v_1$. Clearly, $v_2 = (1, 2, 1)$ will do. Finally, we have to choose $v_3 \in \operatorname{Ker}(f - 2 \operatorname{id})$ with $\{v_1, v_2, v_3\}$ independent. Clearly, $v_3 = (1, 1, 0)$ will do. Thus a Jordan basis is

$$B = \{(1, 1, 1), (1, 2, 1), (1, 1, 0)\}.$$

The transition matrix from B to the standard basis is

$$P = \begin{bmatrix} 1 & 1 & 1 \\ 1 & 2 & 1 \\ 1 & 1 & 0 \end{bmatrix}.$$

We invite the reader to verify that $P^{-1}AP = J$.

An interesting consequence of the Jordan form is the following result.

4.6 Theorem *Every square matrix A over $\mathbb{C}$ is similar to its transpose.*

Proof Because of the form of the Jordan canonical matrix, it clearly suffices to establish the result when A is an elementary Jordan matrix of the form

$$A = \begin{bmatrix} & 1 & & & \\ & & 1 & & \\ & & & \ddots & \\ & & & & 1 \\ & & & & \end{bmatrix}.$$

Now if

$$B = \{v_1, \ldots, v_k\}$$

is an associated Jordan basis define

$$(i = 1, \ldots, k) \qquad w_i = v_{k-i+1}$$

and consider the ordered basis

$$B^* = \{w_1, \ldots, w_k\} = \{v_k, \ldots, v_1\}.$$

Now it is readily seen that the matrix relative to this basis is A^t. Consequently we have that A is similar to A^t. ◊

We shall now illustrate the usefulness of the Jordan form in solving systems of linear differential equations. It is not our intention to become heavily involved with the theory. A little by way of explanation together with some illustrative examples is all we have in mind.

By a *system of linear differential equations* with constant coefficients we mean a system of equations of the form

$$\begin{aligned} x_1' &= a_{11}x_1 + a_{12}x_2 + \cdots + a_{1n}x_n \\ x_2' &= a_{21}x_1 + a_{22}x_2 + \cdots + a_{2n}x_n \\ &\vdots \\ x_n' &= a_{n1}x_1 + a_{n2}x_2 + \cdots + a_{nn}x_n \end{aligned}$$

where $x_1, \ldots, x_n$ are real functions, x_i' denotes the derivative of x_i, and $a_{ij} \in \mathbb{R}$ for all i, j. These equations can be written in the matrix form

$$\tag{1} \mathbf{X}' = A\mathbf{X}$$

where $\mathbf{X} = [x_1 \ \ldots \ x_n]^t \in \text{Mat}_{n \times 1}(\mathbb{R})$ and $A = [a_{ij}]_{n \times n}$. Suppose that A can be reduced to Jordan normal form J_A, and let P be an invertible matrix such that $P^{-1}AP = J_A$. Writing $\mathbf{Y} = P^{-1}\mathbf{X}$, we have

$$\tag{2} (P\mathbf{Y})' = \mathbf{X}' = A\mathbf{X} = AP\mathbf{Y}$$

and so

$$\tag{3} \mathbf{Y}' = P^{-1}\mathbf{X}' = P^{-1}AP\mathbf{Y} = J_A \mathbf{Y}.$$

Now the form of J_A means that (3) is a system that is considerably easier to solve for $\mathbf{Y}$; then, by (2), $P\mathbf{Y}$ is a solution of (1).

Example Consider the system

$$\begin{aligned} x_1' &= x_1 + x_2 \\ x_2' &= -x_1 + 3x_2 \\ x_3' &= -x_1 + 4x_2 - x_3 \end{aligned}$$

i.e. $\mathbf{X}' = A\mathbf{X}$ where

$$\mathbf{X} = \begin{bmatrix} x_1 \\ x_2 \\ x_3 \end{bmatrix}, \qquad A = \begin{bmatrix} 1 & 1 & 0 \\ -1 & 3 & 0 \\ -1 & 4 & -1 \end{bmatrix}.$$

We have $\chi_A = (X+1)(X-2)^2 = m_A$ and so the Jordan form of A is

$$J_A = \begin{bmatrix} -1 & & \\ & 2 & 1 \\ & & 2 \end{bmatrix}.$$

We now determine an invertible matrix P such that $P^{-1}AP = J_A$. For this, we determine a Jordan basis. Let us do so with

REDUCTION TO JORDAN FORM

matrices rather than mappings, for a change. Clearly, we have to find independent column vectors $\mathbf{p}_1, \mathbf{p}_2, \mathbf{p}_3$ such that

$$(A + I_3)\mathbf{p}_1 = \mathbf{0},$$
$$(A - 2I_3)\mathbf{p}_2 = \mathbf{0},$$
$$(A - 2I_3)\mathbf{p}_3 = \mathbf{p}_2.$$

Suitable vectors are, for example,

$$\mathbf{p}_1 = \begin{bmatrix} 0 \\ 0 \\ 1 \end{bmatrix}, \quad \mathbf{p}_2 = \begin{bmatrix} 1 \\ 1 \\ 1 \end{bmatrix}, \quad \mathbf{p}_3 = \begin{bmatrix} -1 \\ 0 \\ 0 \end{bmatrix}.$$

Thus we can take

$$P = \begin{bmatrix} 0 & 1 & -1 \\ 0 & 1 & 0 \\ 1 & 1 & 0 \end{bmatrix}.$$

(Check that $P^{-1}AP = J_A$ or, equivalently, that $AP = PJ_A$.)

With $\mathbf{Y} = P^{-1}\mathbf{X}$ we now solve $\mathbf{Y}' = J_A \mathbf{Y}$, i.e.

$$y_1' = -y_1,$$
$$y_2' = 2y_2 + y_3,$$
$$y_3' = 2y_3.$$

The first and third of these equations give $y_1 = \alpha_1 e^{-t}$ and $y_3 = \alpha_3 e^{2t}$, and the second equation becomes

$$y_2' = 2y_2 + \alpha_3 e^{2t}$$

so that $y_2 = \alpha_3 t e^{2t} + \alpha_2 e^{2t}$. Consequently we see that

$$\mathbf{Y} = \begin{bmatrix} \alpha_1 e^{-t} \\ \alpha_2 e^{2t} + \alpha_3 t e^{2t} \\ \alpha_3 e^{2t} \end{bmatrix}.$$

A solution of the original system of equations is then given by

$$\mathbf{X} = P\mathbf{Y} = \begin{bmatrix} \alpha_2 e^{2t} + \alpha_3(t-1)e^{2t} \\ \alpha_2 e^{2t} + \alpha_3 t e^{2t} \\ \alpha_1 e^{-t} + \alpha_2 e^{2t} + \alpha_3 t e^{2t} \end{bmatrix}.$$

CHAPTER FIVE

The rational and classical forms

Although in general the minimum polynomial of a linear mapping $f : V \to V$ can be expressed as a product of powers of irreducible polynomials over the ground field F of V, say

$$m_f = p_1^{e_1} p_2^{e_2} \ldots p_k^{e_k},$$

the irreducible polynomials p_i need not be linear. Put another way, the eigenvalues of f need not in general all lie in the ground field F. It is natural, therefore, to seek a canonical matrix representation for f in the general case, which will reduce to the Jordan representation when all the eigenvalues of f do belong to F.

In order to develop the machinery to deal with this, we first consider the following notion.

Suppose that W is a subspace of the vector space V. Then in particular W is a (normal) subgroup of the additive group of V and so we can form the quotient group V/W. The elements of this are the cosets

$$x + W = \{x + w \; ; \; w \in W\},$$

and the group operation is given by

$$(x + W) + (y + W) = (x + y) + W.$$

Now we can define a multiplication by scalars on V/W by setting

$$\lambda(x + W) = \lambda x + W.$$

With respect to this, it is readily seen that V/W becomes a vector space over F. We call this the *quotient space* of V by W and denote it also by V/W.

5.1 Theorem *If V is a finite-dimensional vector space and W is a subspace of V then the quotient space V/W is also finite-dimensional. Moreover, if $\{v_1,\ldots,v_m\}$ is a basis of W and $\{x_1+W,\ldots,x_k+W\}$ is a basis of V/W then*

$$B = \{v_1,\ldots,v_m, x_1,\ldots,x_k\}$$

is a basis of V.

Proof The natural mapping $\natural : V \to V/W$ is given by $\natural(x) = x + W$ and is linear. In fact,

$$\natural(x+y) = x+y+W = (x+W)+(y+W) = \natural(x)+\natural(y);$$
$$\natural(\lambda x) = \lambda x + W = \lambda(x+W) = \lambda\natural(x).$$

Suppose now that $\{x_1+W,\ldots,x_k+W\}$ is any linearly independent subset of V/W. Then the set $\{x_1,\ldots,x_k\}$ of coset representatives is a linearly independent subset of V. For, suppose that $\sum_{i=1}^{k} \lambda_i x_i = 0_V$. Then, using the linearity of $\natural$, we have

$$0_{V/W} = \natural(0_V) = \natural\left(\sum_{i=1}^{k} \lambda_i x_i\right) = \sum_{i=1}^{k} \lambda_i \natural(x_i) = \sum_{i=1}^{k} \lambda_i (x_i + W)$$

and so each $\lambda_i = 0$. Consequently $k \leq \dim V$ and V/W is of finite dimension.

Consider now the set B. Applying $\natural$ to any linear combination of elements of B we see as above that B is linearly independent. Now for every $x \in V$ we have $\natural(x) \in V/W$ so there exist scalars λ_i such that

$$x + W = \sum_{i=1}^{k} \lambda_i(x_i + W) = \left(\sum_{i=1}^{k} \lambda_i x_i\right) + W$$

and hence $x - \sum_{i=1}^{k} \lambda_i x_i \in W$, so that $x - \sum_{i=1}^{k} \lambda_i x_i = \sum_{j=1}^{m} \mu_j v_j$. Thus B also spans V and hence is a basis. $\diamond$

5.2 Corollary $\dim V = \dim W + \dim V/W$. $\diamond$

5.3 Corollary *If $V = W \oplus Z$ then $Z \simeq V/W$.*

Proof We have $\dim V = \dim W + \dim Z$ so, by 5.2, $\dim Z = \dim V/W$ and it follows that $Z \simeq V/W$. $\diamond$

We shall be particularly interested in the quotient space V/W when W is a subspace that is f-invariant. In this situation we have the following result.

5.4 Theorem *Let V be a finite-dimensional vector space and let $f : V \to V$ be linear. If W is an f-invariant subspace of V then the prescription*

$$f'(x + W) = f(x) + W$$

defines a linear mapping $f' : V/W \to V/W$, the minimum polynomial of which divides the minimum polynomial of f.

Proof Observe that if $x + W = y + W$ then $x - y \in W$ and so, since W is f-invariant,

$$f(x) - f(y) = f(x - y) \in W$$

which gives $f(x) + W = f(y) + W$. Thus f' indeed defines a mapping from V/W to itself. To see that f' is linear we observe that

$$\begin{aligned} f'[(x + W) + (y + W)] &= f(x + y) + W \\ &= f(x) + f(y) + W \\ &= [f(x) + W] + [f(y) + W]; \end{aligned}$$

$$f'[\lambda(x + W)] = f(\lambda x) + W = \lambda f(x) + W = \lambda[f(x) + W].$$

Now for all positive integers n we have $(f^n)' = (f')^n$. This is readily seen by induction. For the anchor point $n = 1$ the result is trivial; and for the inductive step we have

$$\begin{aligned} (f^{n+1})'(x + W) &= f^{n+1}(x) + W \\ &= f[f^n(x)] + W \\ &= f'[f^n(x) + W] \\ &= f'[(f')^n(x + W)] \\ &= (f')^{n+1}(x + W). \end{aligned}$$

Thus, for every polynomial $p = \sum_{i=1}^{n} a_i X^i$ we have $[p(f)]' = p(f')$. Consequently, taking $p = m_f$ we see that $0 = m_f(f')$ and hence that $m_{f'} | m_f$. $\diamond$

Definition We call $f': V/W \to V/W$ the *linear mapping induced by f on the quotient space* V/W.

We shall now consider a particular type of f-invariant subspace. Let x be a non-zero element of V and consider the set Z_x of all elements of V of the form $p(f)(x)$ where p ranges over all polynomials in $F[X]$. It is clear that Z_x is a subspace of V, and that it is f-invariant.

Example Let $f : \mathbb{R}^3 \to \mathbb{R}^3$ be given by

$$f(x,y,z) = (-y+z, x+z, 2z).$$

Consider the element $(1,0,0)$. We have $f(1,0,0) = (0,1,0)$ and $f^2(1,0,0) = f(0,1,0) = -(1,0,0)$, from which it follows that

$$Z_{(1,0,0)} = \{(x,y,0) \; ; \; x,y \in \mathbb{R}\}.$$

Our immediate objective is to discover a basis for the subspace Z_x. For this purpose, consider the sequence

$$x, f(x), f^2(x), \ldots, f^r(x), \ldots$$

of elements of Z_x. Clearly, there exists a least positive integer k such that $f^k(x)$ is a linear combination of the elements that precede it in this list, say

$$f^k(x) = \lambda_0 x + \lambda_1 f(x) + \cdots + \lambda_{k-1} f^{k-1}(x),$$

and $\{x, f(x), \ldots, f^{k-1}(x)\}$ is then a linearly independent subset of Z_x. Writing $a_i = -\lambda_i$ for $i = 0, \ldots, k-1$ we deduce that the polynomial

$$m_x = a_0 + a_1 X + \cdots + a_{k-1} X^{k-1} + X^k$$

is the monic polynomial of least degree such that $m_x(f)$ 'annihilates' x, in the sense that $m_x(f)(x) = 0_V$.

Definition We call m_x the *f-annihilator* of x.

Example Referring to the previous Example, let $x = (1,0,0)$. Then we have $f^2(x) = -x$. It follows that the f-annihilator of x is $m_x = X^2 + 1$.

With the above notation, we have the following result.

5.5 Theorem *If $x \in V$ has f-annihilator*

$$m_x = a_0 + a_1 X + \cdots + a_{k-1} X^{k-1} + X^k$$

then the set $B_x = \{x, f(x), \ldots, f^{k-1}(x)\}$ is a basis of Z_x, so that $\dim Z_x = \deg m_x$. Moreover, if $f_x : Z_x \to Z_x$ is the induced linear mapping on the f-invariant subspace Z_x then the matrix of f_x relative to the basis B_x is

$$C_{m_x} = \begin{bmatrix} 0 & 0 & 0 & \cdots & 0 & -a_0 \\ 1 & 0 & 0 & \cdots & 0 & -a_1 \\ 0 & 1 & 0 & \cdots & 0 & -a_2 \\ \vdots & \vdots & \vdots & & \vdots & \vdots \\ 0 & 0 & 0 & \cdots & 1 & -a_{k-1} \end{bmatrix}.$$

Finally, the minimum polynomial of f_x is m_x.

Proof Clearly, B_x is linearly independent and $f^k(x) \in \langle B_x \rangle$. We prove by induction that $f^n(x) \in \langle B_x \rangle$ for every n. This is clear for $n = 1, \ldots, k$. Suppose then that $n > k$ and that $f^{n-1}(x) \in \langle B_x \rangle$. Then $f^{n-1}(x)$ is a linear combination of $x, f(x), \ldots, f^{k-1}(x)$ and so $f^n(x) = f[f^{n-1}(x)]$ is a linear combination of $f(x), f^2(x), \ldots, f^k(x)$. Since $f^k(x) \in \langle B_x \rangle$ it follows that $f^n(x) \in \langle B_x \rangle$. It is immediate from this observation that $p(f)(x) \in \langle B_x \rangle$ for every polynomial p. Thus $Z_x \subseteq \langle B_x \rangle$ whence we have equality, the reverse inclusion being obvious. It now follows that B_x is a basis of Z_x.

Since

$$f_x(x) = f(x)$$
$$f_x[f(x)] = f^2(x)$$
$$\vdots$$
$$f_x[f^{k-2}(x)] = f^{k-1}(x)$$
$$f_x[f^{k-1}(x)] = f^k(x) = -a_0 x - a_1 f(x) - \cdots - a_{k-1} f^{k-1}(x)$$

it is clear that the matrix of f_x relative to the basis B_x is the above matrix C_{m_x}.

Finally, suppose that the minimum polynomial of f_x is

$$m_{f_x} = b_0 + b_1 X + \cdots + b_{r-1} X^{r-1} + X^r.$$

Then we have

$$0_V = m_{f_x}(f_x)(x) = m_{f_x}(f)(x) = b_0 x + \cdots + b_{r-1} f^{r-1}(x) + f^r(x)$$

from which $f^r(x)$ is a linear combination of $x, f(x), \ldots, f^{r-1}(x)$ and therefore $k \leq r$. But $m_x(f)$ is the zero map on Z_x, whence so is $m_x(f_x)$. Consequently we have $m_{f_x} | m_x$ and so $r \leq k$. Thus $r = k$ and it follows that $m_{f_x} = m_x$. $\diamond$

Definition We shall call Z_x the *f-cyclic subspace spanned by* $\{x\}$, and C_{m_x} the *companion matrix* of the f-annihilator m_x. Any basis of the form B_x will be called a *cyclic basis*, and x will be called a *cyclic vector*. A subspace that has a cyclic basis will be called a *cyclic subspace*.

Our first main objective can now be revealed. It is to prove that if $f : V \to V$ has minimum polynomial of the form p^k where p is irreducible then V can be expressed as a direct sum of f-cyclic subspaces, the main consequence of this being that f then has a block diagonal representation by companion matrices. Before establishing these facts, we require the following observation.

5.6 Theorem *Let W be an f-invariant subspace of V. Then both the f-annihilator of x and the f'-annihilator of $x+W$ divide the minimum polynomial of f.*

Proof By 5.5, the f-annihilator of x is the minimum polynomial of f_x, the mapping induced on Z_x by f, which clearly divides the minimum polynomial of f.

As for the f'-annihilator of $x + W$, this likewise divides the minimum polynomial of f' which, by 5.4, divides that of f. $\diamond$

5.7 Theorem [Cyclic Decomposition] *Let V be a non-zero vector space of finite dimension and let $f : V \to V$ be linear with minimum polynomial $m_f = p^t$ where p is irreducible. Then there are cyclic vectors $x_1, \ldots, x_k$ and positive integers $n_1, \ldots, n_k$ with each $n_i \leq t$ such that*

(1) $V = \bigoplus_{i=1}^{k} Z_{x_i}$;

(2) *the f-annihilator of x_i is p^{n_i}.*

Proof The proof is by induction on $\dim V$. If $\dim V = 1$ the result is trivial. Suppose then that the result holds for all vector spaces of dimension less than n (where $n > 1$) and let V be of dimension n.

As $m_f = p^t$, there is a non-zero $x_1 \in V$ with $p^{t-1}(f)(x_1) \neq 0_V$. The f-annihilator of x_1 is then $m_{x_1} = p^t$. Let $W = Z_{x_1}$ and let $f' : V/W \to V/W$ be the induced mapping. By 5.4, the minimum polynomial of f' divides $m_f = p^t$ and so the inductive hypothesis applies to f' and V/W. Thus there exist f'-cyclic subspaces $Z_{y_2+W}, \ldots, Z_{y_k+W}$ of V/W such that

$$V/W = \bigoplus_{i=2}^{k} Z_{y_i+W}$$

and, for $2 \leq i \leq k$, the f'-annihilator of $y_i + W$ is p^{n_i} for some $n_i \leq t$.

We now observe that there exists $x_i \in y_i + W$ such that the f-annihilator of x_i is p^{n_i}. In fact, since the f'-annihilator of $y_i + W$ is p^{n_i}, we have $p(f)^{n_i}(y_i) \in W = Z_{x_1}$. Thus there is a polynomial h such that

$$p(f)^{n_i}(y_i) = h(f)(x_1).$$

It follows from this that

$$0_V = p(f)^t(y_i) = p(f)^{t-n_i} h(f)(x_1).$$

But p^t is the f-annihilator of x_1, so $p^t | p^{t-n_i} h$ and hence $p^{n_i} | h$. Consequently $h = p^{n_i} q$ for some polynomial q. Now define $x_i = y_i - q(f)(x_1)$. Then

$$y_i - x_i = q(f)(x_1) \in W$$

and so $x_i \in y_i + W$. The f'-annihilator of $y_i + W$ therefore divides the f-annihilator of x_i. But

$$\begin{aligned} p(f)^{n_i}(x_i) &= p(f)^{n_i}[y_i - q(f)(x_1)] \\ &= p(f)^{n_i}(y_i) - h(f)(x_1) \\ &= 0_V. \end{aligned}$$

Thus we see that the f-annihilator of x_i is p^{n_i}.

Suppose now that $\deg p = d$. Then $\deg p^{n_i} = dn_i$. Since p^{n_i} is both the f-annihilator of x_i and the f'-annihilator of $x_i + W$, it follows by 5.5 that

$$A_i = \{x_i, f(x_i), \ldots, f^{dn_i-1}(x_i)\}$$

is a basis for Z_{x_i}, and that

$$B_i = \{x_i + W, f'(x_i + W), \ldots, (f')^{dn_i-1}(x_i + W)\}$$

is a basis for Z_{x_i+W}. But since

$$V/W = \bigoplus_{i=2}^{k} Z_{y_i+W} = \bigoplus_{i=2}^{k} Z_{x_i+W}$$

it follows that $\bigcup_{i=2}^{k} B_i$ is a basis for V/W. Then, by 5.1, $\bigcup_{i=1}^{k} A_i$ is a basis for V. Consequently we have that

$$V = \bigoplus_{i=1}^{k} Z_{x_i}. \quad \diamond$$

5.8 Corollary *With the above notation, relative to the basis $\bigcup_{i=1}^{k} A_i$ the matrix of f is of the form*

$$\bigoplus_{i=1}^{k} C_i = \begin{bmatrix} C_1 & & & \\ & C_2 & & \\ & & \ddots & \\ & & & C_k \end{bmatrix}$$

where C_i is the companion matrix of $m_{x_i} = p^{n_i}$. $\diamond$

5.9 Corollary $\dim V = (n_1 + \cdots + n_k) \deg p.$ $\diamond$

Without loss of generality, we can assume that the cyclic vectors $x_1, \ldots, x_k$ of 5.7 are arranged such that the corresponding integers n_i satisfy

$$t = n_1 \geq n_2 \geq \cdots \geq n_k \geq 1.$$

With this convention, we have :

5.10 Theorem $n_1, \ldots, n_k$ *are uniquely determined by* f.

Proof From the above we have, for every i,
$$\dim Z_{x_i} = \deg m_{x_i} = \deg p^{n_i} = dn_i.$$

Observe that for every integer j the image of Z_{x_i} under $p(f)^j$ is the f-cyclic subspace $Z_{p(f)^j(x_i)}$. Since the f-annihilator of x_i is p^{n_i}, of degree dn_i, we see that the dimension of $Z_{p(f)^j(x_i)}$ is 0 if $j \geq n_i$ and is $d(n_i - j)$ if $j < n_i$.

Now every $x \in V$ can be written uniquely in the form
$$x = v_1 + \cdots + v_k \quad (v_i \in Z_{x_i})$$
and so every element of $\operatorname{Im} p(f)^j$ can be written uniquely in the form
$$p(f)^j(x) = p(f)^j(v_1) + \cdots + p(f)^j(v_k).$$
Thus, if r is the integer such that $n_1, \ldots, n_r > j$ and $n_{r+1} \leq j$ we see that
$$\operatorname{Im} p(f)^j = \bigoplus_{i=1}^{r} Z_{p(f)^j(x_i)}$$
and consequently
$$\dim \operatorname{Im} p(f)^j = d \sum_{i=1}^{r} (n_i - j) = d \sum_{n_i > j} (n_i - j).$$
It follows from this that
$$\dim \operatorname{Im} p(f)^{j-1} - \dim \operatorname{Im} p(f)^j$$
$$= d\bigg(\sum_{n_i > j-1} (n_i - j + 1) - \sum_{n_i > j} (n_i - j) \bigg)$$
$$= d \sum_{n_i \geq j} (n_i - j + 1 - n_i + j)$$
$$= d \sum_{n_i \geq j} 1$$
$$= d \times \text{number of } n_i \geq j.$$

Now the dimensions on the left are determined by f so the above expression gives, for each j, the number of n_i that are greater than or equal to j. This determines the sequence
$$t = n_1 \geq n_2 \geq \cdots \geq n_k \geq 1$$
completely. $\diamond$

THE RATIONAL AND CLASSICAL FORMS

Definition If the minimum polynomial of f is of the form p^t where p is irreducible then, relative to the uniquely determined chain of integers $t = n_1 \geq n_2 \geq \cdots \geq n_k \geq 1$, the polynomials $p^t = p^{n_1}, p^{n_2}, \ldots, p^{n_k}$ are called the *elementary divisors* of f.

It should be noted that the first elementary divisor in the sequence is the minimum polynomial of f.

We can now apply the above results to the general situation where the characteristic and minimum polynomials of a linear mapping $f : V \to V$ are

$$\chi_f = p_1^{d_1} p_2^{d_2} \ldots p_k^{d_k}, \quad m_f = p_1^{e_1} p_2^{e_2} \ldots p_k^{e_k}$$

where $p_1, \ldots, p_k$ are distinct irreducible polynomials. We know by the Primary Decompositon Theorem that there is a basis of V with respect to which the matrix of f is a block diagonal matrix

$$\begin{bmatrix} A_1 & & & \\ & A_2 & & \\ & & \ddots & \\ & & & A_k \end{bmatrix}$$

in which A_i is the matrix (of size $d_i \deg p_i \times d_i \deg p_i$) that represents the induced mapping f_i on $V_i = \operatorname{Ker} p_i(f)^{e_i}$. Now the minimum polynomial of f_i is $p_i^{e_i}$ and so, by the Cyclic Decomposition Theorem, there is a basis of V_i with respect to which A_i is the block diagonal matrix

$$\begin{bmatrix} C_{i1} & & & \\ & C_{i2} & & \\ & & \ddots & \\ & & & C_{it} \end{bmatrix}$$

in which the C_{ij} are the companion matrices associated with the elementary divisors of f_i. By the previous discussion, this block diagonal form, in which each block A_i is itself a block diagonal of companion matrices, is unique (to within the order of the A_i). It is called the *rational canonical matrix* of f.

It is important to note that in the sequence of elementary divisors there can be repetitions, for some of the n_i can be equal. The result of this is that some companion matrices can appear more than once in the rational form.

Example Suppose that $f : \mathbb{R}^4 \to \mathbb{R}^4$ has minimum polynomial
$$m_f = X^2 + 1.$$
Then $\chi_f = (X^2 + 1)^2$. By 5.9 we have $4 = (n_1 + \cdots + n_k)2$. Since the first elementary divisor is the minimum polynomial, we must have $n_1 = 1$. Since we must also have each $n_i \geq 1$, it follows that the only possibility is $k = 2$ with $n_1 = n_2 = 1$. The rational canonical matrix of f is therefore

$$C_{X^2+1} \oplus C_{X^2+1} = \begin{bmatrix} 0 & -1 & & \\ 1 & 0 & & \\ & & 0 & -1 \\ & & 1 & 0 \end{bmatrix}.$$

Example Suppose now that $f : \mathbb{R}^6 \to \mathbb{R}^6$ has minimum polynomial
$$m_f = (X^2 + 1)(X - 2)^2.$$
The characteristic polynomial of f is then one of
$$\chi_1 = (X^2 + 1)^2(X - 2)^2, \quad \chi_2 = (X^2 + 1)(X - 2)^4.$$
Suppose first that $\chi_f = \chi_1$. Then, arguing exactly as in the previous Example, we see that the rational canonical matrix is
$$C_{X^2+1} \oplus C_{X^2+1} \oplus C_{(X-2)^2}.$$
Suppose now that $\chi_f = \chi_2$. In this case we know that $\mathbb{R}^6 = V_1 \oplus V_2$ with $\dim V_1 = 2$ and $\dim V_2 = 4$. Also, the induced mapping f_2 on V_2 has minimum polynomial $(X - 2)^2$. By 5.9 applied to $f_2 : V_2 \to V_2$ we have $4 = n_1 + \cdots + n_k$ with $n_1 = 2$. There are therefore two possibilities, namely
$$k = 2 \text{ with } n_1 = n_2 = 2;$$
$$k = 3 \text{ with } n_1 = 2, n_2 = n_3 = 1.$$
The rational canonical matrix of f is therefore of one of the forms
$$C_{X^2+1} \oplus C_{(X-2)^2} \oplus C_{(X-2)^2},$$
$$C_{X^2+1} \oplus C_{(X-2)^2} \oplus C_{X-2} \oplus C_{X-2}.$$

Note from the above Example that a knowledge of both the characteristic and the minimum polynomials is not in general enough to determine completely the rational form.

THE RATIONAL AND CLASSICAL FORMS 55

Note also that the rational form is quite different from the Jordan form. To see this, let us take a matrix in Jordan form and find its rational form.

Example Consider the matrix

$$A = \begin{bmatrix} 2 & 1 & 0 \\ 0 & 2 & 1 \\ 0 & 0 & 2 \end{bmatrix}.$$

We have $\chi_A = (X-2)^3 = m_A$ and, by 5.9, $3 = n_1 + \cdots + n_k$ with $n_1 = 3$. Thus $k = 1$ and the rational form is

$$C_{(X-2)^3} = \begin{bmatrix} 0 & 0 & 8 \\ 1 & 0 & -12 \\ 0 & 1 & 6 \end{bmatrix}.$$

The fact that the rational form is quite different from the Jordan form suggests that we are not yet finished, for what we want is a general canonical form that will reduce to the Jordan form when the eigenvalues lie in the ground field. We shall now obtain such a form by modifying the cyclic bases used to obtain the rational form. In so doing, we shall obtain a matrix representation constructed from the companion matrix of p_i rather than those of $p_i^{n_i}$.

5.11 Theorem *Let x be a cyclic vector of V and let $f : V \to V$ have minimum polynomial p^n where*

$$p = \alpha_0 + \alpha_1 X + \cdots + \alpha_{k-1} X^{k-1} + X^k.$$

Then there is a basis of V with respect to which the matrix of f is the $kn \times kn$ matrix

$$\begin{bmatrix} C_p & M & & & \\ & C_p & M & & \\ & & \ddots & \ddots & \\ & & & C_p & M \\ & & & & C_p \end{bmatrix}$$

in which C_p is the companion matrix of p, and M is the $k \times k$ matrix
$$\begin{bmatrix} & & 1 \\ & & \\ & & \end{bmatrix}.$$

Proof Consider the kn elements
$$\begin{array}{ccc} f^{k-1}(x), & \ldots, & f(x), \quad x \\ p(f)[f^{k-1}(x)], & \ldots, & p(f)[f(x)], \quad p(f)(x) \\ \vdots & & \vdots \quad\quad \vdots \\ p(f)^{n-1}[f^{k-1}(x)], \ldots, & p(f)^{n-1}[f(x)], p(f)^{n-1}(x). \end{array}$$

To show that this set is a basis of V it suffices to show that it is linearly independent. Suppose that it were not so. Then some non-trivial linear combination of these elements would be 0_V and so there would exist a polynomial h such that $h(f)(x) = 0_V$ with $\deg h < kn = \deg p^n$. Since x is cyclic, this contradicts the assumption that p^n is the minimum polynomial of f. We order this basis in a row-by-row manner, as we normally read.

Now f maps each element in the above array to its predecessor in the same row, except those at the beginning of a row. For these elements we have, for example,
$$\begin{aligned} f[f^{k-1}(x)] &= f^k(x) \\ &= -\alpha_{k-1} f^{k-1}(x) - \cdots - \alpha_0 x + p(f)(x). \end{aligned}$$

It is now an easy matter to verify that the matrix of f relative to the above basis is of the form described. ◇

Definition A block matrix of the form described in 5.11 will be called a *classical p-matrix* associated with the companion matrix C_p.

Applying 5.11 to the cyclic subspaces appearing in the Cyclic Decomposition Theorem, we see that in the rational canonical matrix of f we can replace each diagonal block of companion matrices associated with the elementary divisors $p_i^{n_i}$ by a classical p_i-matrix associated with the companion matrix of p_i. This gives another canonical matrix which we call the *classical canonical matrix* of f.

Example Let $f : \mathbb{R}^6 \to \mathbb{R}^6$ be such that

$$\chi_f = m_f = (X^2 - X + 1)^2 (X + 1)^2.$$

Then the rational canonical matrix of f is

$$C_{(X^2-X+1)^2} \oplus C_{(X+1)^2} = \begin{bmatrix} 0 & 0 & 0 & -1 & & \\ 1 & 0 & 0 & 2 & & \\ 0 & 1 & 0 & -3 & & \\ 0 & 0 & 1 & 2 & & \\ & & & & 0 & -1 \\ & & & & 1 & -2 \end{bmatrix}$$

and the classical canonical matrix is

$$\begin{bmatrix} 0 & -1 & 0 & 1 & & \\ 1 & 1 & 0 & 0 & & \\ & & 0 & -1 & & \\ & & 1 & 1 & & \\ & & & & -1 & 1 \\ & & & & & -1 \end{bmatrix}.$$

Finally, let us note that if in 5.11 we have $p = X - \alpha$ (so that $k = 1$ and $f - \alpha \,\mathrm{id}_V$ is nilpotent of index n) then C_p is the 1×1 matrix $[\alpha]$ and the classical p-matrix associated with C_p reduces to the $n \times n$ elementary Jordan matrix associated with the eigenvalue α. Thus the classical form reduces to the Jordan form when the eigenvalues belong to the ground field.

CHAPTER SIX

Dual spaces

If V and W are vector spaces over a field F then the set $\text{Lin}(V,W)$ of linear mappings from V to W is also a vector space over F : if $f,g \in \text{Lin}(V,W)$ define $f+g$ and λf by $(f+g)(x) = f(x)+g(x)$ and $(\lambda f)(x) = \lambda f(x)$, and observe that $f+g, \lambda f$ belong to $\text{Lin}(V,W)$. A particular case of this is of especial importance, namely that in which for W we take the ground field F (regarded as a vector space over itself). It is on this vector space $\text{Lin}(V,F)$ that we shall now focus our attention.

Definition By the *dual space* of V we shall mean the vector space $\text{Lin}(V,F)$, which we shall denote by V^d. The elements of V^d, i.e. the linear mappings $f : V \to F$, will be called *linear functionals* (or *linear forms*) on V.

Example The i-th projection p_i given by $p_i(x_1,\ldots,x_n) = x_i$ is a linear functional on $\mathbb{R}^n$ so is an element of $(\mathbb{R}^n)^d$.

Example If $V = \text{Mat}_{n \times n}(\mathbb{C})$ then $T : V \to \mathbb{C}$ given by $T(A) = \sum_{i=1}^{n} a_{ii}$ is a linear functional on V so is an element of V^d.

Example The mapping $I : \mathbb{R}[X] \to \mathbb{R}$ given by $I(p) = \int_0^1 p$ is a linear functional on $\mathbb{R}[X]$ so is an element of $\mathbb{R}[X]^d$.

In what follows, we shall denote a typical element of V^d by x^d. Thus the notation x^d will be used to denote a linear mapping from V to the ground field F.

We begin by showing that if V is of finite dimension then so is the dual space V^d. This we do by constructing a basis of V^d from a basis for V.

DUAL SPACES

6.1 Theorem Let $\{v_1, \ldots, v_n\}$ be a basis of V and for $i = 1, \ldots, n$ let $v_i^d : V \to F$ be the linear mapping such that

$$v_i^d(v_j) = \begin{cases} 1_F & \text{if } j = i; \\ 0_F & \text{if } j \neq i. \end{cases}$$

Then $\{v_1^d, \ldots, v_n^d\}$ is a basis of V^d.

Proof It is clear that $v_i^d \in V^d$. Suppose that $\sum_{i=1}^{n} \lambda_i v_i^d = 0$ in V^d. Then for $j = 1, \ldots, n$ we have

$$0_F = \left(\sum_{i=1}^{n} \lambda_i v_i^d\right)(v_j) = \sum_{i=1}^{n} \lambda_i v_i^d(v_j) = \sum_{i=1}^{n} \lambda_i \delta_{ij} = \lambda_j$$

and so $\{v_1^d, \ldots, v_n^d\}$ is linearly independent. If $x = \sum_{j=1}^{n} x_j v_j \in V$ then we have

$$(\star) \qquad v_i^d(x) = \sum_{j=1}^{n} x_j v_i^d(v_j) = \sum_{j=1}^{n} x_j \delta_{ij} = x_i$$

and hence, for any $f \in V^d$,

$$\left(\sum_{i=1}^{n} f(v_i) v_i^d\right)(x) = \sum_{i=1}^{n} f(v_i) v_i^d(x)$$
$$= \sum_{i=1}^{n} f(v_i) x_i = f\left(\sum_{i=1}^{n} x_i v_i\right) = f(x).$$

Thus we see that

$$(\star\star) \qquad (\forall f \in V^d) \qquad f = \sum_{i=1}^{n} f(v_i) v_i^d$$

which shows that $\{v_1^d, \ldots, v_n^d\}$ also spans V^d, whence it is a basis. $\diamond$

6.2 Corollary If $\dim V$ is finite then $\dim V^d = \dim V$. $\diamond$

Note from $(\star)$ and $(\star\star)$ in the above proof that

$$(\forall x \in V) \qquad x = \sum_{i=1}^{n} v_i^d(x) v_i;$$

$$(\forall x^d \in V^d) \qquad x^d = \sum_{i=1}^{n} x^d(v_i)v_i^d.$$

Definition If $\{v_1, \ldots, v_n\}$ is a basis of V then we shall say that the basis $\{v_1^d, \ldots, v_n^d\}$ of V^d described in 6.1 is the corresponding *dual basis*.

Because of $(\star)$ above, the mappings $v_1^d, \ldots, v_n^d$ are often called the *coordinate forms* associated with $v_1, \ldots, v_n$.

Example Consider the basis $\{v_1, v_2\}$ of $\mathbb{R}^2$ where $v_1 = (1,2)$ and $v_2 = (2,3)$. Let $\{v_1^d, v_2^d\}$ be the dual basis. Then we have

$$1 = v_1^d(v_1) = v_1^d(1,2) = v_1^d(1,0) + 2v_1^d(0,1);$$
$$0 = v_1^d(v_2) = v_1^d(2,3) = 2v_1^d(1,0) + 3v_1^d(0,1).$$

These equations give $v_1^d(1,0) = -3$ and $v_1^d(0,1) = 2$ and hence v_1^d is given by

$$v_1^d(x,y) = -3x + 2y.$$

Similarly, we have

$$v_2^d(x,y) = 2x - y.$$

Example Consider the standard basis $\{e_1, \ldots, e_n\}$ of $\mathbb{R}^n$. By definition, we have $e_i^d(e_j) = \delta_{ij}$ and so

$$e_i^d(x_1, \ldots, x_n) = e_i^d\left(\sum_{j=1}^{n} x_j e_j\right) = \sum_{j=1}^{n} x_j e_i^d(e_j) = x_i$$

whence the dual basis is the set of projections $\{p_1, \ldots, p_n\}$.

Example Let $t_1, \ldots, t_{n+1}$ be $n+1$ distinct real numbers and for each i let $\varsigma_{t_i} : \mathbb{R}_n[X] \to \mathbb{R}$ be the substitution mapping given by $\varsigma_{t_i}(p) = p(t_i)$. Then

$$B = \{\varsigma_{t_1}, \ldots, \varsigma_{t_{n+1}}\}$$

is a basis for $\mathbb{R}_n[X]^d$. In fact, since $\mathbb{R}_n[X]^d$ has the same dimension as $\mathbb{R}_n[X]$, namely $n+1$, it suffices to prove that B is

DUAL SPACES

linearly independent. But if $\sum_{i=1}^{n+1} \lambda_i \varsigma_{t_i} = 0$ then

$$0 = \Big(\sum_{i=1}^{n+1} \lambda_i \varsigma_{t_i}\Big)(1) = \lambda_1 + \lambda_2 + \cdots + \lambda_{n+1}$$

$$0 = \Big(\sum_{i=1}^{n+1} \lambda_i \varsigma_{t_i}\Big)(X) = \lambda_1 t_1 + \lambda_2 t_2 + \cdots + \lambda_{n+1} t_{n+1}$$

$$\vdots$$

$$0 = \Big(\sum_{i=1}^{n+1} \lambda_i \varsigma_{t_i}\Big)(X^n) = \lambda_1 t_1^n + \lambda_2 t_2^n + \cdots + \lambda_{n+1} t_{n+1}^n.$$

The coefficient matrix of this system of equations is the *Vandermonde matrix*

$$M = \begin{bmatrix} 1 & 1 & \cdots & 1 \\ t_1 & t_2 & \cdots & t_{n+1} \\ \vdots & \vdots & & \vdots \\ t_1^n & t_2^n & \cdots & t_{n+1}^n \end{bmatrix}.$$

By induction, it can be shown that $\det M = \prod_{j<i}(t_i - t_j)$. Since $t_1, \ldots, t_{n+1}$ are distinct, it follows that $\det M \neq 0$ whence we see that $\lambda_1 = \cdots = \lambda_{n+1} = 0$.

To find a basis of $\mathbb{R}_n[X]$ of which B is the dual, let such a basis be

$$A = \{p_1, \ldots, p_{n+1}\}.$$

Then we require $\varsigma_{t_i}(p_j) = \delta_{ij}$ for all i, j. It is readily seen that the *Lagrange polynomials*

$$L_j = \prod_{i \neq j} \frac{X - t_i}{t_j - t_i}$$

are the successful candidates.

6.3 Theorem *Let $(v_i)_n, (w_i)_n$ be ordered bases of V and let $(v_i^d)_n, (w_i^d)_n$ be the corresponding dual bases. If P is the transition matrix from $(v_i)_n$ to $(w_i)_n$ then the transition matrix from $(v_i^d)_n$ to $(w_i^d)_n$ is $(P^{-1})^t$.*

Proof Let Q be the transition matrix from $(v_i^d)_n$ to $(w_i^d)_n$. Then we have

$$w_i = \sum_{j=1}^n p_{ji} v_j, \qquad w_i^d = \sum_{j=1}^n q_{ji} v_j^d.$$

Consequently,

$$\begin{aligned}
\delta_{ij} = w_i^d(w_j) &= \sum_{k=1}^n q_{ki} \left(\sum_{t=1}^n p_{tj} v_k^d(v_t) \right) \\
&= \sum_{k=1}^n q_{ki} \left(\sum_{t=1}^n p_{tj} \delta_{kt} \right) \\
&= \sum_{k=1}^n q_{ki} p_{kj} \\
&= [Q^t P]_{ij}.
\end{aligned}$$

Thus we see that $Q^t P = I_n$ and so $Q = (P^{-1})^t$. $\diamond$

The result of 6.3 provides a useful way of exhibiting dual bases. Given a basis

$$B = \{(a_{11}, \ldots, a_{1n}), (a_{21}, \ldots, a_{2n}), \ldots, (a_{n1}, \ldots, a_{nn})\}$$

of $\mathbb{R}^n$, the transition matrix from B to the standard basis $(e_i)_n$ is the matrix P whose i-th *column* is $[a_{i1} \ \ldots \ a_{in}]^t$. By 6.3, if we denote the i-th *row* of P^{-1} by $[\alpha_{i1} \ \ldots \ \alpha_{in}]$ and if we define the mapping $[\alpha_{i1}, \ldots, \alpha_{in}]: \mathbb{R}^n \to \mathbb{R}$ by

$$[\alpha_{i1}, \ldots, \alpha_{in}](x_1, \ldots, x_n) = \alpha_{i1} x_1 + \cdots + \alpha_{in} x_n,$$

then we see that

$$[\alpha_{i1}, \ldots, \alpha_{in}](a_{j1}, \ldots, a_{jn}) = [P^{-1} P]_{ij} = \delta_{ij}$$

and hence we can represent the dual basis of B by

$$B^\star = \{[\alpha_{11}, \ldots, \alpha_{1n}], [\alpha_{21}, \ldots, \alpha_{2n}], \ldots, [\alpha_{n1}, \ldots, \alpha_{nn}]\}.$$

Example Consider again the basis $B = \{(1,2), (2,3)\}$ of $\mathbb{R}^2$. The transition matrix from B to the standard basis is

$$P = \begin{bmatrix} 1 & 2 \\ 2 & 3 \end{bmatrix}$$

DUAL SPACES

and its inverse is
$$P^{-1} = \begin{bmatrix} -3 & 2 \\ 2 & -1 \end{bmatrix}.$$
The basis dual to B can be described as $B^\star = \{[-3,2],[2,-1]\}$ where
$$[-3,2](x,y) = -3x + 2y, \qquad [2,-1](x,y) = 2x - y.$$

Example Consider again the standard basis $B = (e_i)_n$ of $\mathbb{R}^n$. Applying the above procedure, we have $P = I_n = P^{-1}$ from which it follows that the dual basis consists of the projections $p_1, \ldots, p_n$.

We now introduce the following notation. Given $x \in V$ and $y^d \in V^d$ we shall write
$$y^d(x) = \langle x, y^d \rangle.$$
With this notation, the following identities are immediate from the linearity of the mappings involved :

$(\alpha) \quad \langle x+y, z^d \rangle = \langle x, z^d \rangle + \langle y, z^d \rangle;$

$(\beta) \quad \langle x, y^d + z^d \rangle = \langle x, y^d \rangle + \langle x, z^d \rangle;$

$(\gamma) \quad \langle \lambda x, y^d \rangle = \lambda \langle x, y^d \rangle;$

$(\delta) \quad \langle x, \lambda y^d \rangle = \lambda \langle x, y^d \rangle.$

Now it is clear from (β) and (δ) that for every $x \in V$ the mapping $\widehat{x} : V^d \to F$ given by
$$\widehat{x}(y^d) = \langle x, y^d \rangle$$
is linear, i.e. is an element of the dual space $\widehat{V} = (V^d)^d$ of V^d. Moreover, by (α) and (γ) it is quickly verified that the mapping $\alpha_V : V \to \widehat{V}$ given by $\alpha_V(x) = \widehat{x}$ is also linear.

Definition We call $\widehat{V}$ the *bidual* of V, and α_V the *canonical map* from V to $\widehat{V}$.

Note that the various notations employed above can be summarised by the following identities :
$$\boxed{\widehat{x}(y^d) = \langle x, y^d \rangle = y^d(x).}$$

6.4 Theorem *If V is of finite dimension then the canonical map $\alpha_V : V \to \widehat{V}$ is an isomorphism.*

Proof Let $\{v_1, \ldots, v_n\}$ be a basis of V and $x = \sum_{i=1}^{n} x_i v_i \in \operatorname{Ker} \alpha_V$. Then $\widehat{x}$ is the zero element of $\widehat{V}$ and so $\widehat{x}(y^d) = 0$ for all $y^d \in V^d$. In particular, for $i = 1, \ldots, n$ we have

$$0 = \widehat{x}(v_i^d) = \langle x, v_i^d \rangle = v_i^d(x) = x_i$$

and consequently $x = 0_V$. Thus α_V is injective. Since, by 6.2,

$$\dim V = \dim V^d = \dim \widehat{V}$$

it follows that α_V is an isomorphism. $\diamondsuit$

In the case where V is of finite dimension we shall agree to *identify* V and $\widehat{V}$. We can do so only because the isomorphism α_V is *natural* in the sense that it is independent of the choice of basis.

Example Let $V = P_1(\mathbb{R})$ be the vector space of polynomials over $\mathbb{R}$ of the form $a_0 + a_1 X$. Then $f_1, f_2 : V \to \mathbb{R}$ given by $f_1(p) = \int_0^1 p$ and $f_2(p) = \int_0^2 p$ are linear functionals on V and $\{f_1, f_2\}$ is a basis of V^d. Now $P_1(\mathbb{R}) \simeq \mathbb{R}^2$ under the assignment

$$a_0 + a_1 X \longleftrightarrow (a_0, a_1)$$

so $P_1(\mathbb{R})^d \simeq (\mathbb{R}^2)^d$. If $p = a_0 + a_1 X$ then

$$f_1(p) = a_0 + \tfrac{1}{2} a_1$$
$$f_2(p) = 2a_0 + 2a_1$$

and so we can associate with $\{f_1, f_2\}$ the basis $\{[1, \tfrac{1}{2}], [2, 2]\}$ of $(\mathbb{R}^2)^d$. By considering the matrix

$$P = \begin{bmatrix} 1 & 2 \\ \tfrac{1}{2} & 2 \end{bmatrix}$$

whose inverse is

$$P^{-1} = \begin{bmatrix} 2 & -2 \\ -\tfrac{1}{2} & 1 \end{bmatrix}$$

we see that $\{(2, -2), (-\tfrac{1}{2}, 1)\}$ is the basis of $\widehat{\mathbb{R}^2} = \mathbb{R}^2$ that is dual to B. Thus $\{2 - 2X, \tfrac{1}{2} + X\}$ is a basis of $P_1(\mathbb{R})$ that is dual to the basis $\{f_1, f_2\}$.

DUAL SPACES

Suppose now that V, W are vector spaces over F and that $f : V \to W$ is linear. Given any $y^d \in W^d = \text{Lin}(W, F)$, consider the linear mapping $y^d \circ f$. The diagram
$$V \xrightarrow{f} W \xrightarrow{y^d} F$$
shows that $y^d \circ f \in V^d$. Thus the assignment $y^d \mapsto y^d \circ f$ defines a mapping from $W^d \to V^d$. We call this the *transpose* of the linear mapping f and denote it by f^t. Thus we have
$$f^t(y^d) = y^d \circ f$$
so f^t can be described as 'composing on the right by f'. Note that then $f^t \in \text{Lin}(W^d, V^d)$; for
$$f^t(y^d + z^d) = (y^d \circ f) + (z^d \circ f) = f^t(y^d) + f^t(z^d);$$
$$f^t(\lambda y^d) = (\lambda y^d) \circ f = \lambda(y^d \circ f) = \lambda[f^t(y^d)].$$
In terms of the notation introduced previously, we have the following identity:
$$\boxed{\langle f(x), y^d \rangle = \langle x, f^t(y^d) \rangle.}$$
In fact,
$$\langle f(x), y^d \rangle = y^d[f(x)] = (y^d \circ f)(x) = [f^t(y^d)](x) = \langle x, f^t(y^d) \rangle.$$
The use of the word 'transpose' is suggested by the following result.

6.5 Theorem *Let V, W be vector spaces of dimensions m, n over F. Let $(a_i)_m, (b_i)_n$ be ordered bases of V, W and let $f : V \to W$ be linear. If the matrix of f relative to $(a_i)_m, (b_i)_n$ is $A = [a_{ij}]_{n \times m}$ then relative to the corresponding dual bases $(a_i^d)_m, (b_i^d)_n$ the matrix of f^t is A^t.*

Proof If the matrix representing f^t is $B = [b_{ij}]_{m \times n}$ then we have
$$\langle f(a_i), b_j^d \rangle = \left\langle \sum_{t=1}^{n} a_{ti} b_t, b_j^d \right\rangle = \sum_{t=1}^{n} a_{ti} \langle b_t, b_j^d \rangle = a_{ji}$$
$$\|$$
$$\langle a_i, f^t(b_j^d) \rangle = \left\langle a_i, \sum_{t=1}^{n} b_{tj} a_t^d \right\rangle = \sum_{t=1}^{n} b_{tj} \langle a_i, a_t^d \rangle = b_{ij}$$
from which we see that $b_{ij} = a_{ji}$ for all i, j and so $B = A^t$. $\diamond$

The principal properties of transposition are listed in the following result.

6.6 Theorem (1) *The transpose of* id_V *is* id_{V^d};
(2) *If* $f, g \in \mathrm{Lin}(V, W)$ *then* $(f+g)^t = f^t + g^t$;
(3) *If* $f \in \mathrm{Lin}(V, W)$ *and* $g \in \mathrm{Lin}(W, X)$ *then* $(g \circ f)^t = f^t \circ g^t$.

Proof All three are immediate from the fact that for every linear mapping h the effect of h^t can be described as composition on the right by h. $\diamond$

6.7 Corollary *If* $f : V \to W$ *is an isomorphism then so is* $f^t : W^d \to V^d$; *moreover, we have* $(f^t)^{-1} = (f^{-1})^t$.

Proof This follows from (1) and (3) on taking $g = f^{-1}$. $\diamond$

Of course, when V and W are finite-dimensional 6.6 and 6.7 follow immediately from 6.5 and the corresponding properties of transposition for matrices.

We can also consider the transpose of f^t. We denote this by f^{tt} and call it the *bitranspose* of f. The connection between bitransposes and biduals is the following.

6.8 Theorem *For every linear mapping* $f : V \to W$ *the diagram*

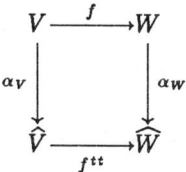

is commutative, in the sense that $f^{tt} \circ \alpha_V = \alpha_W \circ f$.

Proof We have to show that $f^{tt}(\widehat{x}) = \widehat{f(x)}$ for every $x \in V$. Now for all $y^d \in V^d$ we have

$$[f^{tt}(\widehat{x})](y^d) = (\widehat{x} \circ f^t)(y^d) = \widehat{x}[f^t(y^d)]$$
$$= \langle x, f^t(y^d) \rangle$$
$$= \langle f(x), y^d \rangle = \widehat{f(x)}(y^d),$$

from which the result follows. $\diamond$

An immediate consequence of 6.8 is that when V and W are of finite dimensions (in which case we agree to identify $V, \widehat{V}$ and $W, \widehat{W}$ and therefore also α_V, id_V and α_W, id_W) we have $f^{tt} = f$. This then matches the matrix situation, where $A^{tt} = A$.

Definition If $x \in V$ and $y^d \in V^d$ are such that $\langle x, y^d \rangle = 0$ then we say that x is *annihilated* by y^d.

Since $\langle x, y^d \rangle = y^d(x)$ we see that the set of elements of V that are annihilated by y^d is $\operatorname{Ker} y^d$. Now it is immediate from the identities $(\beta), (\gamma)$ preceding 6.4 that, for every non-empty subset E of V, the set of elements of V^d that annihilate every element of E is a subspace of V^d. We denote this subspace by E°. Thus

$$E^\circ = \{y^d \in V^d \,;\, (\forall x \in E) \quad \langle x, y^d \rangle = 0\}.$$

We call E° the *annihilator* of E. It is clear that $\{0_V\}^\circ = V^d$ and that $V^\circ = \{0_{V^d}\}$.

6.9 Theorem *If V is a finite-dimensional vector space and W is a subspace of V then*

$$\dim W^\circ = \dim V - \dim W.$$

Moreover, identifying V and $\widehat{V}$, we have $W = W^{\circ\circ}$.

Proof Let $\dim V = n$. The result is trivial if $W = V$ so suppose that $W \subset V$. Then $\dim W = m < n$. Let $\{a_1, \ldots, a_m\}$ be a basis of W and extend this to a basis

$$\{a_1, \ldots, a_m, a_{m+1}, \ldots, a_n\}$$

of V. Let $\{a_1^d, \ldots, a_n^d\}$ be the corresponding dual basis. If $x^d = \sum_{i=1}^{n} \lambda_i a_i^d \in W^\circ$ then for $j = 1, \ldots, m$ we have

$$0 = \langle a_j, x^d \rangle = \sum_{i=1}^{n} \lambda_i \langle a_j, a_i^d \rangle = \lambda_j.$$

It follows that $\{a_{m+1}^d, \ldots, a_n^d\}$ is a basis of W° and consequently

$$\dim W^\circ = n - m = \dim V - \dim W.$$

As for the second statement, consider the subspace $W^{\circ\circ} = (W^\circ)^\circ$ of $\widehat{V} = V$. By definition, every element of W is annihilated by every element of W° and so we have $W \subseteq W^{\circ\circ}$. On the other hand, by what we have just proved,

$$\dim W^{\circ\circ} = n - \dim W^\circ = n - (n - m) = m = \dim W.$$

It follows, therefore, that $W = W^{\circ\circ}$. $\diamond$

Annihilators and transposes are connected :

6.10 Theorem *If V, W are finite-dimensional and $f : V \to W$ is linear then*
(1) $(\operatorname{Im} f)^\circ = \operatorname{Ker} f^t$;
(2) $(\operatorname{Ker} f)^\circ = \operatorname{Im} f^t$;
(3) $\dim \operatorname{Im} f^t = \dim \operatorname{Im} f$;
(4) $\dim \operatorname{Ker} f^t = \dim \operatorname{Ker} f$.

Proof (1) We have $y^d \in (\operatorname{Im} f)^\circ$ if and only if, for every $x \in V$,

$$0 = \langle f(x), y^d \rangle = \langle x, f^t(y^d) \rangle$$

which is the case if and only if $f^t(y^d) \in V^\circ = \{0_V\}$, i.e. if and only if $y^d \in \operatorname{Ker} f^t$.

(2) Replacing f by f^t in (1) and using the fact that $f^{tt} = f$, we obtain $(\operatorname{Im} f^t)^\circ = \operatorname{Ker} f$. Then, by 6.9, $(\operatorname{Ker} f)^\circ = (\operatorname{Im} f^t)^{\circ\circ} = \operatorname{Im} f^t$.

(3),(4) follow from (1),(2), and 6.9. $\diamondsuit$

6.11 Corollary *The row and column rank of a matrix A over a field F are the same.*

Proof If A represents a linear mapping f then A^t represents f^t. The result follows from the fact that the row rank of A is $\dim \operatorname{Im} f$ and the column rank of A is the row rank of A^t which is $\dim \operatorname{Im} f^t$. $\diamondsuit$

CHAPTER SEVEN

Inner product spaces

In some aspects of our discussion of vector spaces the ground field F has played no significant rôle. In this Chapter we shall restrict F to be $\mathbb{R}$ or $\mathbb{C}$, the results we obtain depending heavily on the properties of these fields.

Definition Let V be a vector space over $\mathbb{C}$. By an *inner product* on V we shall mean a mapping $f : V \times V \to \mathbb{C}$, described by $(x, y) \mapsto \langle x\,|\,y\rangle$, such that for all $x, x', y \in V$ and all $\alpha \in \mathbb{C}$ the following identities hold:

(1) $\quad \langle x + x'\,|\,y\rangle = \langle x\,|\,y\rangle + \langle x'\,|\,y\rangle$;

(2) $\quad \langle \alpha x\,|\,y\rangle = \alpha\langle x\,|\,y\rangle$;

(3) $\quad \overline{\langle x\,|\,y\rangle} = \langle y\,|\,x\rangle$, so that in particular $\langle x\,|\,x\rangle \in \mathbb{R}$;

(4) $\quad \langle x\,|\,x\rangle \geq 0$, with equality if and only if $x = 0_V$.

By a *complex inner product space* we mean a vector space V over $\mathbb{C}$ together with an inner product on V. By a *real inner product space* we mean a vector space V over $\mathbb{R}$ together with an inner product on V (this being defined as in the above but with the bar denoting complex conjugate omitted). By an *inner product space* we shall mean either a complex inner product space or a real inner product space.

There are certain other identities that follow immediately from (1) to (4) above, namely:

(5) $\quad \langle x\,|\,y + y'\rangle = \langle x\,|\,y\rangle + \langle x\,|\,y'\rangle$.

In fact, by (1) and (3) we have

$$\langle x\,|\,y + y'\rangle = \overline{\langle y + y'\,|\,x\rangle} = \overline{\langle y\,|\,x\rangle} + \overline{\langle y'\,|\,x\rangle} = \langle x\,|\,y\rangle + \langle x\,|\,y'\rangle.$$

(6) $\langle x \mid \alpha y \rangle = \overline{\alpha}\langle x \mid y \rangle$.

This follows from (3) and (4) since

$$\langle x \mid \alpha y \rangle = \overline{\langle \alpha y \mid x \rangle} = \overline{\alpha \langle y \mid x \rangle} = \overline{\alpha}\,\overline{\langle y \mid x \rangle} = \overline{\alpha}\langle x \mid y \rangle.$$

(7) $\langle x \mid 0 \rangle = 0 = \langle 0 \mid x \rangle$.

This is immediate from (1), (2), (3) on taking $x' = -x, y' = -y$, and $\alpha = -1$.

Example $\mathbb{C}^n$ is a complex inner product space under the mapping described by $(z, w) \mapsto \langle z \mid w \rangle$ where

$$\langle (z_1, \ldots, z_n) \mid (w_1, \ldots, w_n) \rangle = \sum_{i=1}^{n} z_i \overline{w_i}.$$

This inner product is called the *standard inner product* on $\mathbb{C}$.

Example $\mathbb{R}^n$ is a real inner product space under the corresponding *standard inner product* given by

$$\langle (x_1, \ldots, x_n) \mid (y_1, \ldots, y_n) \rangle = \sum_{i=1}^{n} x_i y_i.$$

In the cases where $n = 2, 3$ this inner product is often called the *dot product* or *scalar product*. This terminology is popular when dealing with the geometric application of vectors. Indeed, several of the results that we shall establish will generalise familiar results in euclidean geometry of two and three dimensions.

Example Let $a, b \in \mathbb{R}$ and let V be the real vector space of continuous functions $f : [a, b] \to \mathbb{R}$. Define a mapping from $V \times V$ to $\mathbb{R}$ by

$$(f, g) \mapsto \langle f \mid g \rangle = \int_a^b fg.$$

Then this defines an inner product on V.

Example Let $\mathbb{R}_n[X]$ be the real vector space of polynomials of degree less than n. Then

$$\langle p \mid q \rangle = \int_0^1 pq$$

defines an inner product on $\mathbb{R}_n[X]$.

Example For an $n \times n$ matrix $A = [a_{ij}]$ let $\operatorname{tr} A = \sum_{i=1}^{n} a_{ii}$. Then the vector space $\operatorname{Mat}_{n \times n}(\mathbb{R})$ can be made into a real inner product space by defining
$$\langle A \mid B \rangle = \operatorname{tr}(B^t A).$$
Likewise, $\operatorname{Mat}_{n \times n}(\mathbb{C})$ can be made into a complex inner product space by defining
$$\langle A \mid B \rangle = \operatorname{tr}(B^\star A)$$
where $B^\star = \overline{B^t}$ is the complex conjugate of the transpose of B.

Definition Let V be an inner product space. For every $x \in V$ we define the *norm* of x to be the non-negative real number $\|x\| = \sqrt{\langle x \mid x \rangle}$. Given $x, y \in V$ we define the *distance between x and y* to be $d(x, y) = \|x - y\|$.

Example In the real inner product space $\mathbb{R}^2$ under the standard inner product, if $x = (x_1, x_2)$ then $\|x\|^2 = x_1^2 + x_2^2$, so $\|x\|$ is the distance from x to the origin. Likewise, if $y = (y_1, y_2)$ then we have $\|x - y\|^2 = (x_1 - y_1)^2 + (x_2 - y_2)^2$, which shows the connection between the general concept of distance and the theorem of Pythagoras.

It is clear from (4) above that $\|x\| = 0$ if and only if $x = 0_V$.

7.1 Theorem *Let V be an inner product space. Then, for all $x, y \in V$ and every scalar λ,*
(1) $\|\lambda x\| = |\lambda| \|x\|$;
(2) [Cauchy-Schwarz inequality] $|\langle x \mid y \rangle| \leq \|x\| \|y\|$;
(3) [Triangle inequality] $\|x + y\| \leq \|x\| + \|y\|$.

Proof (1) $\|\lambda x\|^2 = \langle \lambda x \mid \lambda x \rangle = \lambda \overline{\lambda} \langle x \mid x \rangle = |\lambda|^2 \|x\|^2$.

(2) The result is trivial if $x = 0_V$. Suppose then that $x \neq 0_V$, so that $\|x\| \neq 0$. Let $z = y - \dfrac{\langle y \mid x \rangle}{\|x\|^2} x$. Then, noting that $\langle z \mid x \rangle = 0$, we have
$$\begin{aligned} 0 \leq \|z\|^2 &= \left\langle y - \frac{\langle y \mid x \rangle}{\|x\|^2} x \;\middle|\; y - \frac{\langle y \mid x \rangle}{\|x\|^2} x \right\rangle \\ &= \langle y \mid y \rangle - \frac{\langle y \mid x \rangle}{\|x\|^2} \langle x \mid y \rangle \\ &= \langle y \mid y \rangle - \frac{|\langle x \mid y \rangle|^2}{\|x\|^2} \end{aligned}$$

from which (2) follows.

(3) This follows from the observation that

$$\begin{aligned}\|x+y\|^2 &= \langle x+y \,|\, x+y\rangle \\ &= \langle x\,|\,x\rangle + \langle x\,|\,y\rangle + \langle y\,|\,x\rangle + \langle y\,|\,y\rangle \\ &= \|x\|^2 + \langle x\,|\,y\rangle + \overline{\langle x\,|\,y\rangle} + \|y\|^2 \\ &= \|x\|^2 + 2\,\mathrm{Re}\langle x\,|\,y\rangle + \|y\|^2 \\ &\le \|x\|^2 + 2|\langle x\,|\,y\rangle| + \|y\|^2 \\ &\le \|x\|^2 + 2\|x\|\,\|y\| + \|y\|^2 \qquad \text{by (2)} \\ &= \bigl(\|x\| + \|y\|\bigr)^2. \qquad \diamondsuit\end{aligned}$$

Example Let V be the set of infinite sequences $(a_i)_{i\ge 1}$ of real numbers that are *square summable* in the sense that $\sum_{i\ge 1} a_i^2$ exists. Defining an addition and a multiplication by real scalars in the obvious component-wise manner, we see that V becomes a real vector space. Let $(a_i)_{i\ge 1}$ and $(b_i)_{i\ge 1}$ be elements of V. By the Cauchy-Schwarz inequality applied to the inner product space $\mathbb{R}^n$ with the standard inner product, we have

$$\Bigl|\sum_{i=1}^{k} a_i b_i\Bigr| \le \sum_{i=1}^{k} a_i^2 \sum_{i=1}^{k} b_i^2 \le \sum_{i\ge 1} a_i^2 \sum_{i\ge 1} b_i^2$$

so the sequence with k-th term $\sum_{i=1}^{k} a_i b_i$ is absolutely summable and hence is summable. Thus $\sum_{i\ge 1} a_i b_i$ exists and we can define

$$\langle (a_i)_{i\ge 1} \,|\, (b_i)_{i\ge 1}\rangle = \sum_{i\ge 1} a_i b_i.$$

In this way, V becomes a real inner product space that is often called ℓ_2-*space* or *Hilbert space*.

Definition If V is an inner product space then $x,y \in V$ are said to be *orthogonal* if $\langle x\,|\,y\rangle = 0$. A non-empty subset S of V is said to be an *orthogonal subset* of V if every pair of distinct elements of S is orthogonal. An *orthonormal subset* of V is an orthogonal subset S such that $\|x\| = 1$ for every $x \in S$, i.e. a set of mutually orthogonal vectors of length 1.

INNER PRODUCT SPACES 73

Example Relative to the standard inner products, the standard bases of $\mathbb{R}^n$ and of $\mathbb{C}^n$ are orthonormal subsets.

Example In $\mathbb{R}^2$ the elements $x = (x_1, x_2)$ and $y = (y_1, y_2)$ are orthogonal if and only if $x_1 y_1 + x_2 y_2 = 0$. Geometrically, this is equivalent to saying that the lines joining x and y to the origin are mutually perpendicular.

Example In the vector space V of real continuous functions on the interval $[-\pi, \pi]$ with inner product $\langle f \mid g \rangle = \int_{-\pi}^{\pi} fg$ the set

$$S = \{x \mapsto 1, x \mapsto \sin kx, x \mapsto \cos kx \; ; \; k = 1, 2, 3, \ldots\}$$

is an orthogonal subset.

It is clear that an orthonormal subset of V can always be obtained from an orthogonal subset S by *normalising* each element x of S, i.e. by replacing x by $x^\star = x/\|x\|$.

An important property of orthogonal (and hence of orthonormal) sets is the following.

7.2 Theorem *Orthogonal sets are linearly independent.*

Proof Let S be an orthogonal subset of V and let $x_1, \ldots, x_n \in S$ be such that $\sum\limits_{i=1}^{n} \lambda_i x_i = 0_V$. Then for every i we have

$$\lambda_i \langle x_i \mid x_i \rangle = \sum_{k=1}^{n} \lambda_k \langle x_k \mid x_i \rangle = \left\langle \sum_{k=1}^{n} \lambda_k x_k \; \middle| \; x_i \right\rangle = \langle 0_V \mid x_i \rangle = 0$$

from which it follows by (4) on page 69 that $\lambda_i = 0$. $\diamondsuit$

We now describe properties of the subspace spanned by an orthonormal subset.

7.3 Theorem *Let $\{e_1, \ldots, e_n\}$ be an orthonormal subset of the inner product space V. Then*

[Bessel's inequality] $\qquad (\forall x \in V) \quad \sum\limits_{k=1}^{n} \left| \langle x \mid e_k \rangle \right|^2 \leq \|x\|^2.$

Moreover, if W is the subspace spanned by $\{e_1, \ldots, e_n\}$ then the following statements are equivalent :

(1) $x \in W$;

(2) $\sum_{k=1}^{n} |\langle x\,|\,e_k\rangle|^2 = \|x\|^2$;

(3) $x = \sum_{k=1}^{n} \langle x\,|\,e_k\rangle e_k$;

(4) $(\forall y \in V)\quad \langle x\,|\,y\rangle = \sum_{k=1}^{n} \langle x\,|\,e_k\rangle\langle e_k\,|\,y\rangle$.

Proof Let $z = x - \sum_{k=1}^{n} \langle x\,|\,e_k\rangle e_k$. Then a simple computation gives

$$0 \le \langle z\,|\,z\rangle = \langle x\,|\,x\rangle - \sum_{k=1}^{n} \langle x\,|\,e_k\rangle \overline{\langle x\,|\,e_k\rangle}$$
$$= \|x\|^2 - \sum_{k=1}^{n} |\langle x\,|\,e_k\rangle|^2.$$

(2) $\Rightarrow$ (3) is now immediate since (2) implies that $z = 0_V$.

(3) $\Rightarrow$ (4) : If $x = \sum_{k=1}^{n} \langle x\,|\,e_k\rangle e_k$ then, for all $y \in V$,

$$\langle x\,|\,y\rangle = \Big\langle \sum_{k=1}^{n} \langle x\,|\,e_k\rangle e_k \,\Big|\, y \Big\rangle = \sum_{k=1}^{n} \langle x\,|\,e_k\rangle\langle e_k\,|\,y\rangle.$$

(4) $\Rightarrow$ (2) follows by taking $y = x$ in (4).

(3) $\Rightarrow$ (1) is clear.

(1) $\Rightarrow$ (3) : If $x = \sum_{k=1}^{n} \lambda_k e_k$ then for $j = 1, \ldots, n$ we have

$$\lambda_j = \sum_{k=1}^{n} \lambda_k \langle e_k\,|\,e_j\rangle = \Big\langle \sum_{k=1}^{n} \lambda_k e_k \,\Big|\, e_j \Big\rangle = \langle x\,|\,e_j\rangle. \quad \diamond$$

Definition By an *orthonormal basis* of an inner product space we mean an orthonormal subset that is a basis.

Example The standard bases of $\mathbb{R}^n$ and $\mathbb{C}^n$ are orthonormal.

Example In $\text{Mat}_{n \times n}(\mathbb{C})$ with $\langle A\,|\,B\rangle = \text{tr}\,(B^\star A)$, an orthonormal basis is $\{E_{pq}\ ;\ p, q = 1, \ldots, n\}$ where E_{pq} has a 1 in the (p, q)-th position and 0 elsewhere.

We shall now show that every finite-dimensional inner product space has an orthonormal basis. In so doing, we give a practical method of constructing such a basis.

INNER PRODUCT SPACES

7.4 Theorem [Gram-Schmidt orthonormalisation process] *Let V be an inner product space and for every non-zero $x \in V$ let $x^* = x/\|x\|$. If $\{x_1, \ldots, x_k\}$ is a linearly independent subset of V, define recursively*

$$y_1 = x_1^*;$$
$$y_2 = \bigl(x_2 - \langle x_2 \,|\, y_1\rangle y_1\bigr)^*;$$
$$y_3 = \bigl(x_3 - \langle x_3 \,|\, y_2\rangle y_2 - \langle x_3 \,|\, y_1\rangle y_1\bigr)^*;$$
$$\vdots$$
$$y_k = \Bigl(x_k - \sum_{i=1}^{k-1} \langle x_k \,|\, y_i\rangle y_i\Bigr)^*.$$

Then $\{y_1, \ldots, y_k\}$ is orthonormal and spans the same subspace as $\{x_1, \ldots, x_k\}$.

Proof It is readily seen that $y_i \neq 0_V$ for every i and that y_i is a linear combination of $x_1, \ldots, x_i$. It is also clear that x_i is a linear combination of $y_1, \ldots, y_i$. Thus $\{x_1, \ldots, x_k\}$ and $\{y_1, \ldots, y_k\}$ span the same subspace.

It now suffices to prove that $\{y_1, \ldots, y_k\}$ is an orthogonal subset; and this we do inductively. For $k = 1$ the result is trivial. Suppose that $\{y_1, \ldots, y_{t-1}\}$ is orthogonal where $t > 1$. Then, writing

$$\Bigl\|x_t - \sum_{i=1}^{t-1} \langle x_t \,|\, y_i\rangle y_i\Bigr\| = \alpha_t,$$

we see that

$$\alpha_t y_t = x_t - \sum_{i=1}^{t-1} \langle x_t \,|\, y_i\rangle y_i$$

and so, for $j < t$,

$$\alpha_t \langle y_t \,|\, y_j\rangle = \langle x_t \,|\, y_j\rangle - \sum_{i=1}^{t-1} \langle x_t \,|\, y_i\rangle \langle y_i \,|\, y_j\rangle$$
$$= \langle x_t \,|\, y_j\rangle - \langle x_t \,|\, y_j\rangle$$
$$= 0.$$

Since $\alpha_t \neq 0$ we deduce that $\langle y_t \,|\, y_j\rangle = 0$ for $j < t$. Thus $\{y_1, \ldots, y_t\}$ is orthogonal. $\diamondsuit$

7.5 Corollary *If V is a finite-dimensional inner product space then V has an orthonormal basis.*

Proof Apply the Gram-Schmidt process to a basis of V. $\diamondsuit$

Example Consider the basis $\{x_1, x_2, x_3\}$ of $\mathbb{R}^3$ where $x_1 = (0, 1, 1), x_2 = (1, 0, 1), x_3 = (1, 1, 0)$. In order to apply the Gram-Schmidt process to this basis using the standard inner product, we first let $y_1 = x_1/\|x_1\| = \frac{1}{\sqrt{2}}(0, 1, 1)$. Now

$$\begin{aligned} x_2 - \langle x_2 \mid y_1 \rangle y_1 &= (1, 0, 1) - \tfrac{1}{\sqrt{2}} \langle (1, 0, 1) \mid (0, 1, 1) \rangle \tfrac{1}{\sqrt{2}}(0, 1, 1) \\ &= (1, 0, 1) - \tfrac{1}{2}(0, 1, 1) \\ &= \tfrac{1}{2}(2, -1, 1) \end{aligned}$$

so, normalising this, we take $y_2 = \frac{1}{\sqrt{6}}(2, -1, 1)$.

Note that, by 7.1, we have

$$(\lambda x)^\star = \frac{\lambda x}{\|\lambda x\|} = \frac{\lambda x}{|\lambda|\, \|x\|} = \begin{cases} x^\star & \text{if } \lambda > 0; \\ -x^\star & \text{if } \lambda < 0. \end{cases}$$

This helps avoid unnecessary arithmetic in the calculations.

Similarly,

$$x_3 - \langle x_3 \mid y_2 \rangle y_2 - \langle x_3 \mid y_1 \rangle y_1 = \tfrac{2}{3}(1, 1, -1)$$

and so, normalising this, we take $y_3 = \frac{1}{\sqrt{3}}(1, 1, -1)$. Then an orthonormal basis is

$$\{\tfrac{1}{\sqrt{2}}(0, 1, 1), \tfrac{1}{\sqrt{6}}(2, -1, 1), \tfrac{1}{\sqrt{3}}(1, 1, -1)\}.$$

7.6 Theorem *If $\{e_1, \ldots, e_n\}$ is an orthonormal basis of an inner product space V then*

(1) $(\forall x \in V) \quad x = \sum_{k=1}^{n} \langle x \mid e_k \rangle e_k;$

(2) $(\forall x \in V) \quad \|x\|^2 = \sum_{k=1}^{n} |\langle x \mid e_k \rangle|^2;$

(3) $(\forall x, y \in V) \quad \langle x \mid y \rangle = \sum_{k=1}^{n} \langle x \mid e_k \rangle \langle e_k \mid y \rangle.$

Proof This is immediate from 7.3. $\diamondsuit$

INNER PRODUCT SPACES 77

Definition The identity (1) in 7.6 is often called the *Fourier expansion of x relative to the orthonormal basis* $\{e_1,\ldots,e_n\}$, the scalars $\langle x \,|\, e_k\rangle$ being called the *Fourier coefficients* of x. The identity (3) of 7.6 is called *Parseval's identity*.

Example The Fourier coefficients of $(1,1,1)$ relative to the orthonormal basis obtained in the previous Example are readily seen to be $\sqrt{2}, \sqrt{2}/\sqrt{3}, 1/\sqrt{3}$.

Just as a linearly independent subset can be extended to a basis, so can an orthonormal subset be extended to an orthonormal basis. This is the content of the following result.

7.7 Theorem *Let V be an inner product space of dimension n. If $\{x_1,\ldots,x_k\}$ is an orthonormal subset of V then there exist $x_{k+1},\ldots,x_n \in V$ such that $\{x_1,\ldots,x_n\}$ is an orthonormal basis of V.*

Proof Let W be the subspace spanned by $\{x_1,\ldots,x_k\}$. By 7.2, this set is a basis for W and so can be extended to a basis

$$\{x_1,\ldots,x_k,z_{k+1},\ldots,z_n\}$$

of V. Applying the Gram-Schmidt process to this basis, we obtain an orthonormal basis of V. The result now follows on noting that the first k terms of this orthonormal basis of V are precisely $x_1,\ldots,x_k$; for, by the formulae in 7.4 and the orthogonality of $x_1,\ldots,x_k$, we see that $y_i = (x_i)^\star = x_i$ for each i. $\diamond$

An isomorphism from one vector space to another carries bases to bases. We shall now investigate the corresponding situation for inner product spaces.

Definition If V, W are inner product spaces over the same field then $f : V \to W$ is called an *inner product isomorphism* if it is a vector space isomorphism that preserves inner products, in the sense that

$$(\forall x, y \in V) \quad \langle f(x) \,|\, f(y)\rangle = \langle x \,|\, y\rangle.$$

7.8 Theorem *Let V, W be inner product spaces over the same field. Let $\{e_1,\ldots,e_n\}$ be an orthonormal basis of V. Then $f : V \to W$ is an inner product isomorphism if and only if $\{f(e_1),\ldots,f(e_n)\}$ is an orthonormal basis of W.*

Proof $\Rightarrow$: If $f : V \to W$ is an inner product isomorphism then clearly $\{f(e_1), \ldots, f(e_n)\}$ is a basis of W. It is also orthonormal since
$$\langle f(e_i) \mid f(e_j) \rangle = \langle e_i \mid e_j \rangle = \begin{cases} 1 & \text{if } i = j; \\ 0 & \text{if } i \neq j. \end{cases}$$

$\Leftarrow$: Suppose now that $\{f(e_1), \ldots, f(e_n)\}$ is an orthonormal basis of W. Then f carries a basis to a basis and so is a vector space isomorphism. Now for all $x \in V$ we have, using the Fourier expansion of x relative to $\{e_1, \ldots, e_n\}$,

$$\begin{aligned} \langle f(x) \mid f(e_j) \rangle &= \Big\langle f\Big(\sum_{i=1}^n \langle x \mid e_i \rangle e_i\Big) \,\Big|\, f(e_j) \Big\rangle \\ &= \Big\langle \sum_{i=1}^n \langle x \mid e_i \rangle f(e_i) \,\Big|\, f(e_j) \Big\rangle \\ &= \sum_{i=1}^n \langle x \mid e_i \rangle \langle f(e_i) \mid f(e_j) \rangle \\ &= \langle x \mid e_j \rangle \end{aligned}$$

and similarly
$$\langle f(e_j) \mid x \rangle = \langle e_j \mid x \rangle.$$

It now follows by Parseval's identity applied to both V and W that
$$\begin{aligned} \langle f(x) \mid f(y) \rangle &= \sum_{j=1}^n \langle f(x) \mid f(e_j) \rangle \langle f(e_j) \mid f(y) \rangle \\ &= \sum_{j=1}^n \langle x \mid e_j \rangle \langle e_j \mid y \rangle \\ &= \langle x \mid y \rangle \end{aligned}$$

and consequently f is an inner product isomorphism. $\diamond$

We now pass to the consideration of the dual of an inner product space. For this purpose, we require the following notions.

Definition Let V, W be vector spaces over a field F where F is either $\mathbb{R}$ or $\mathbb{C}$. A mapping $f : V \to W$ is called a *conjugate transformation* if

$$(\forall x, y \in V)(\forall \lambda \in F) \quad f(x+y) = f(x) + f(y), \quad f(\lambda x) = \overline{\lambda} f(x).$$

If, furthermore, f is a bijection then we say that it is a *conjugate isomorphism*.

INNER PRODUCT SPACES

Note that when $F = \mathbb{R}$ conjugate transformations are simply linear mappings.

We now observe that for every $y \in V$ the mapping from V to F described by $x \mapsto \langle x \mid y \rangle$ is linear, and hence is an element of V^d. We shall write this element as y^d, so that we have the following useful amalgamated notation :

$$\boxed{\langle x \mid y \rangle = y^d(x) = \langle x, y^d \rangle.}$$

7.9 Theorem *If V is a finite-dimensional inner product space then there is a conjugate isomorphism $\vartheta_V : V \to V^d$, namely that given by $\vartheta_V(x) = x^d$ where $(\forall x \in V)\ x^d(y) = \langle y, x^d \rangle$.*

Proof Consider the mapping $\vartheta_V : V \to V^d$ given by $\vartheta_V(x) = x^d$. Since

$$\begin{aligned}\langle x, (y+z)^d \rangle &= \langle x \mid y+z \rangle = \langle x \mid y \rangle + \langle x \mid z \rangle \\ &= \langle x, y^d \rangle + \langle x, z^d \rangle = \langle x, y^d + z^d \rangle\end{aligned}$$

we see that $(y+z)^d = y^d + z^d$, so $\vartheta_V(y+z) = \vartheta_V(y) + \vartheta_V(z)$. Likewise,

$$\langle x, (\lambda y)^d \rangle = \langle x \mid \lambda y \rangle = \overline{\lambda}\langle x \mid y \rangle = \overline{\lambda}\langle x, y^d \rangle = \langle x, \overline{\lambda} y^d \rangle$$

and so $(\lambda y)^d = \overline{\lambda} y^d$ whence $\vartheta_V(\lambda y) = \overline{\lambda}\vartheta(y)$. Thus ϑ_V is a conjugate transformation.

That ϑ_V is injective follows from the fact that if $x \in \operatorname{Ker} \vartheta_V$ then $x^d = 0$ and so $\langle x \mid x \rangle = \langle x, x^d \rangle = 0$ whence $x = 0_V$.

To show that ϑ_V is also surjective, let $f \in V^d$. If $\{e_1, \ldots, e_n\}$ is an orthonormal basis of V, let

$$x = \sum_{i=1}^{n} \overline{f(e_i)} e_i.$$

Then for $j = 1, \ldots, n$ we have

$$x^d(e_j) = \langle e_j \mid x \rangle = \Big\langle e_j \,\Big|\, \sum_{i=1}^{n} \overline{f(e_i)} e_i \Big\rangle = \sum_{i=1}^{n} f(e_i) \langle e_j \mid e_i \rangle = f(e_j).$$

Thus x^d and f coincide on the basis $\{e_1, \ldots, e_n\}$. We deduce that $x^d = \vartheta_V(x)$ and so ϑ_V is also surjective. Thus ϑ_V is a conjugate isomorphism. $\diamondsuit$

We note from the above that we have the identity

$$(\forall x, y \in V) \qquad \langle x \,|\, y \rangle = \langle x, \vartheta_V(y) \rangle.$$

Since ϑ_V is a bijection, we also have the following identity (obtained by writing $\vartheta_V^{-1}(y)$ instead of y):

$$(\forall x, y \in V) \qquad \langle x \,|\, \vartheta_V^{-1}(y) \rangle = \langle x, y \rangle.$$

We can now establish the following important result.

7.10 Theorem *Let V and W be finite-dimensional inner product spaces over the same field. Then for every linear mapping $f : V \to W$ there is a unique linear mapping $f^* : W \to V$ such that*

$$(\forall x \in V)(\forall y \in W) \qquad \langle f(x) \,|\, y \rangle = \langle x \,|\, f^*(y) \rangle.$$

Proof With the above notation, we have the identity

$$\begin{aligned}
\langle f(x) \,|\, y \rangle &= \langle f(x), y^d \rangle = \langle x, f^t(y^d) \rangle \\
&= \langle x \,|\, \vartheta_V^{-1}[f^t(y^d)] \rangle \\
&= \langle x \,|\, (\vartheta_V^{-1} \circ f^t \circ \vartheta_W)(y) \rangle,
\end{aligned}$$

from which it follows immediately that

$$f^* = \vartheta_V^{-1} \circ f^t \circ \vartheta_W$$

is the only linear mapping with the stated property. $\diamond$

7.11 Corollary $f^* : W \to V$ *is the unique linear mapping such that the diagram*

$$\begin{array}{ccc}
W & \xrightarrow{\vartheta_W} & W^d \\
{\scriptstyle f^*}\Big\downarrow & & \Big\downarrow {\scriptstyle f^t} \\
V & \xrightarrow[\vartheta_V]{} & V^d
\end{array}$$

is commutative, in the sense that $\vartheta_V \circ f^* = f^t \circ \vartheta_W$. $\diamond$

Definition The unique linear mapping f^* of 7.10 will be called the *adjoint* of f.

Immediate properties of the assignment $f \mapsto f^*$ are the following.

INNER PRODUCT SPACES

7.12 Theorem *Let V, W, X be finite-dimensional inner product spaces over the same field. Let $f, g : V \to W$ and $h : W \to X$ be linear mappings. Then*

(1) $(f + g)^* = f^* + g^*$;
(2) $(\lambda f)^* = \overline{\lambda} f^*$;
(3) $(h \circ f)^* = f^* \circ h^*$;
(4) $(f^*)^* = f$.

Proof (1) is immediate from $f^* = \vartheta_V^{-1} \circ f^t \circ \vartheta_W$ and the fact that $(f + g)^t = f^t + g^t$.

(2) $\langle (\lambda f)(x) \,|\, y \rangle = \lambda \langle f(x) \,|\, y \rangle = \lambda \langle x \,|\, f^*(y) \rangle = \langle x \,|\, \overline{\lambda} f^*(y) \rangle$ and so, by the uniqueness of adjoints, $(\lambda f)^* = \overline{\lambda} f^*$.

(3) $\langle h[f(x)] \,|\, y \rangle = \langle f(x) \,|\, h^*(y) \rangle = \langle x \,|\, f^*[h^*(y)] \rangle$ and so, by the uniqueness of adjoints, $(h \circ f)^* = f^* \circ h^*$.

(4) Taking complex conjugates in 7.10 we obtain the identity

$$\langle f^*(y) \,|\, x \rangle = \langle y \,|\, f(x) \rangle,$$

from which it follows by the uniqueness of adjoints that $(f^*)^* = f$. $\diamond$

7.13 Theorem *Let V and W be finite-dimensional inner product spaces over the same field with $\dim V = \dim W$. If $f : V \to W$ is linear then the following statements are equivalent :*

(1) *f is an inner product isomorphism*;
(2) *f is a vector space isomorphism and $f^{-1} = f^*$*;
(3) $f \circ f^* = \mathrm{id}_W$;
(4) $f^* \circ f = \mathrm{id}_V$.

Proof $(1) \Rightarrow (2)$: If (1) holds then f^{-1} exists and we have the identity

$$\langle f(x) \,|\, y \rangle = \langle f(x) \,|\, f[f^{-1}(y)] \rangle = \langle x \,|\, f^{-1}(y) \rangle,$$

from which it follows by the uniqueness of adjoints that $f^{-1} = f^*$.

It is clear that $(2) \Rightarrow (3)$ and $(2) \Rightarrow (4)$.

$(4) \Rightarrow (1)$: If (4) holds then f is injective, hence bijective, and $f^{-1} = f^*$. Consequently,

$$(\forall x, y \in V) \qquad \langle f(x) \,|\, f(y) \rangle = \langle x \,|\, f^*[f(y)] \rangle = \langle x \,|\, y \rangle$$

and so f is an inner product isomorphism.

The proof of (3) $\Rightarrow$ (1) is similar. $\diamond$

We have seen in 6.10 how the transpose f^t of f is such that Ker f^t and Im f^t are the annihilators of Im f and Ker f respectively. In view of the connection between transposes and adjoints, it will come as no surprise that Ker f^* and Im f^* are also related to the subspaces Im f and Ker f. This connection is via the following notion.

Definition Let V be an inner product space. For every non-empty subset E of V we define the *orthogonal complement* of E in V to be the set

$$E^\perp = \{y \in V \;;\; (\forall x \in E) \;\; \langle x \,|\, y \rangle = 0\}.$$

It is clear that $E^\perp$ is a subspace of V. The terminology is suggested by the following result.

7.14 Theorem *Let V be an inner product space and let W be a finite-dimensional subspace of V. Then*

$$V = W \oplus W^\perp.$$

Proof Let $\{e_1, \ldots, e_n\}$ be an orthonormal basis of W, noting that this exists since W is of finite dimension. Given $x \in V$, let $x' = \sum_{i=1}^{n} \langle x \,|\, e_i \rangle e_i$ and let $x'' = x - x'$. Then $x' \in W$ and for $j = 1, \ldots, n$ we have

$$\begin{aligned}
\langle x'' \,|\, e_j \rangle &= \langle x \,|\, e_j \rangle - \langle x' \,|\, e_j \rangle \\
&= \langle x \,|\, e_j \rangle - \sum_{i=1}^{n} \langle x \,|\, e_i \rangle \langle e_i \,|\, e_j \rangle \\
&= \langle x \,|\, e_j \rangle - \langle x \,|\, e_j \rangle \\
&= 0.
\end{aligned}$$

It follows that $x'' \in W^\perp$ and hence that $x = x' + x'' \in W + W^\perp$. Thus $V = W + W^\perp$. Now if $x \in W \cap W^\perp$ we have $\langle x \,|\, x \rangle = 0$ whence $\|x\| = 0$ and so $x = 0_V$. Thus we conclude that $V = W \oplus W^\perp$. $\diamond$

INNER PRODUCT SPACES

Example The above result has a basic application to the theory of *Fourier series*. Suppose that W is a finite-dimensional subspace of the inner product space V. Given $x \in V$ let $x = a+b$ where $a \in W$ and $b \in W^\perp$. Then, by orthogonality,

$$\|x\|^2 = \langle a + b \mid a + b \rangle = \|a\|^2 + \|b\|^2.$$

For any $y \in W$ we deduce that

$$\begin{aligned}\|x - y\|^2 = \|a - y + b\|^2 &= \|a - y\|^2 + \|b\|^2 \\ &= \|a - y\|^2 + \|x - a\|^2 \\ &\geq \|x - a\|^2.\end{aligned}$$

Thus we see that the element of W that is 'nearest' the element x of V is the component a of x in W. Now let $\{e_1, \ldots, e_n\}$ be an orthonormal basis for W. Let the element of W that is nearest a given $x \in V$ be the element $a = \sum_{i=1}^{n} \lambda_i e_i$. By 7.3 we have $\lambda_i = \langle a \mid e_i \rangle$ and by orthogonality $\langle x \mid e_i \rangle = \langle a + b \mid e_i \rangle = \langle a \mid e_i \rangle$. Thus the element of W that is nearest x is $\sum_{i=1}^{n} \langle x \mid e_i \rangle e_i$, the scalars being the Fourier coefficients.

Now apply these observations to the inner product space V of continuous functions $f : [-\pi, \pi] \to \mathbb{R}$ under the inner product $\langle f \mid g \rangle = \frac{1}{\pi} \int_{-\pi}^{\pi} fg$. An orthonormal subset of V is

$$S = \{x \mapsto \tfrac{1}{\sqrt{2}}, x \mapsto \sin kx, x \mapsto \cos kx \ ; \ k = 1, 2, 3, \ldots\}.$$

Let W_n be the $(2n+1)$-dimensional subspace spanned by

$$B_n = \{x \mapsto \tfrac{1}{\sqrt{2}}, x \mapsto \sin kx, x \mapsto \cos kx \ ; \ k = 1, \ldots, n\}.$$

Then the element f_n of W_n that is nearest a given $f \in V$ is of the form

$$f_n = \tfrac{1}{2}a_0 + \sum_{k=1}^{n}(a_k \cos kx + b_k \sin kx)$$

where

$$a_0 = \frac{1}{\pi}\int_{-\pi}^{\pi} f(x)\, dx, \quad a_k = \frac{1}{\pi}\int_{-\pi}^{\pi} f(x) \cos kx\, dx,$$

$$b_k = \frac{1}{\pi}\int_{-\pi}^{\pi} f(x) \sin kx\, dx.$$

If f is infinitely differentiable then it can be shown that the sequence $(f_n)_{n\geq 1}$ is a Cauchy sequence having f as its limit. Thus we can write

$$f = \tfrac{1}{2}a_0 + \sum_{k\geq 1}(a_k\cos kx + b_k\sin kx)$$

which is the *Fourier series representation* of f.

7.15 Theorem *If V is a finite-dimensional inner product space and W is a subspace of V then $W = W^{\perp\perp}$ and*

$$\dim W^\perp = \dim V - \dim W.$$

Proof By 7.14 we clearly have $\dim V = \dim W + \dim W^\perp$. Now it is clear from the definition of $W^\perp$ that we have $W \subseteq W^{\perp\perp}$. Also,
$$\dim W^{\perp\perp} = \dim V - \dim W^\perp = \dim W.$$
It follows that $W = W^{\perp\perp}$. $\diamond$

7.16 Theorem *If V is a finite-dimensional inner product space and A, B are subspaces of V then*
(1) $A \subseteq B \Longrightarrow B^\perp \subseteq A^\perp$;
(2) $(A \cap B)^\perp = A^\perp + B^\perp$;
(3) $(A + B)^\perp = A^\perp \cap B^\perp$.

Proof (1) If $A \subseteq B$ then clearly every element that is orthogonal to B is orthogonal to A, so $B^\perp \subseteq A^\perp$.

(2) Since $A, B \subseteq A+B$ we have, by (1), $(A+B)^\perp \subseteq A^\perp \cap B^\perp$; and since $A \cap B \subseteq A, B$ we have $A^\perp, B^\perp \subseteq (A \cap B)^\perp$ whence $A^\perp + B^\perp \subseteq (A \cap B)^\perp$. Since then

$$A \cap B = (A \cap B)^{\perp\perp} \subseteq (A^\perp + B^\perp)^\perp \subseteq A^{\perp\perp} \cap B^{\perp\perp} = A \cap B$$

we deduce that $A \cap B = (A^\perp + B^\perp)^\perp$ whence $(A \cap B)^\perp = A^\perp + B^\perp$.

(3) This follows from (2) on replacing A, B by $A^\perp, B^\perp$. $\diamond$

7.17 Theorem *If V is a finite-dimensional inner product space and if $f : V \to V$ is linear then*

$$\operatorname{Im} f^\star = (\operatorname{Ker} f)^\perp \quad \text{and} \quad \operatorname{Ker} f^\star = (\operatorname{Im} f)^\perp.$$

INNER PRODUCT SPACES

Proof Let $z \in \operatorname{Im} f^*$, say $z = f^*(y)$. Then for every $x \in \operatorname{Ker} f$ we have
$$\langle x \mid z \rangle = \langle x \mid f^*(y) \rangle = \langle f(x) \mid y \rangle = \langle 0_V \mid y \rangle = 0$$
and consequently $z \in (\operatorname{Ker} f)^\perp$. Thus $\operatorname{Im} f^* \subseteq (\operatorname{Ker} f)^\perp$.

Now let $y \in \operatorname{Ker} f^*$. Then for $z = f(x) \in \operatorname{Im} f$ we have
$$\langle z \mid y \rangle = \langle f(x) \mid y \rangle = \langle x \mid f^*(y) \rangle = \langle x \mid 0_V \rangle = 0$$
and consequently $y \in (\operatorname{Im} f)^\perp$. Thus $\operatorname{Ker} f^* \subseteq (\operatorname{Im} f)^\perp$.

Using 7.15 we then have
$$\begin{aligned}
\dim \operatorname{Im} f &= \dim V - \dim(\operatorname{Im} f)^\perp \\
&\leq \dim V - \dim \operatorname{Ker} f^* \\
&= \dim \operatorname{Im} f^* \\
&\leq \dim(\operatorname{Ker} f)^\perp \\
&= \dim V - \dim \operatorname{Ker} f \\
&= \dim \operatorname{Im} f.
\end{aligned}$$

The resulting equality gives both $\dim \operatorname{Im} f^* = \dim(\operatorname{Ker} f)^\perp$ and $\dim(\operatorname{Im} f)^\perp = \dim \operatorname{Ker} f^*$, from which the results follow. $\diamondsuit$

We now investigate how matrices that represent f and f^* are related.

Definition If $A = [a_{ij}]_{m \times n} \in \operatorname{Mat}_{m \times n}(\mathbb{C})$ then by the *adjoint* (or *conjugate transpose*) of A we mean the $n \times m$ matrix A^* such that $[A^*]_{ij} = \overline{a_{ji}}$.

The following result justifies the above terminology.

7.18 Theorem *Let V, W be finite-dimensional inner product spaces over the same field. If, relative to ordered orthonormal bases $(v_i)_n, (w_i)_m$ a linear mapping $f : V \to W$ is represented by the matrix A then the mapping f^* is represented, relative to the bases $(w_i)_m$ and $(v_i)_n$, by the matrix A^*.*

Proof For $j = 1, \ldots, n$ we have $f(v_j) = \sum_{i=1}^{m} \langle f(v_j) \mid w_i \rangle w_i$ by 7.6, so if $A = [a_{ij}]$ we have $a_{ij} = \langle f(v_j) \mid w_i \rangle$. Likewise, we have $f^*(v_j) = \sum_{i=1}^{n} \langle f^*(w_j) \mid v_i \rangle v_i$. Then since
$$\overline{a_{ij}} = \overline{\langle f(v_j) \mid w_i \rangle} = \langle w_i \mid f(v_j) \rangle = \langle f^*(w_i) \mid v_j \rangle$$

it follows that the matrix that represents f^* is A^*. $\diamond$

It is clear from 7.18 and 7.13 that a square matrix A represents an inner product space isomorphism if and only if A^{-1} exists and is A^*. Such a matrix is said to be *unitary*. It is readily seen by extending the corresponding results for ordinary vector spaces to inner product spaces that if A, B are $n \times n$ matrices over the ground field of V then A, B represent the same linear mapping with respect to possibly different ordered orthonormal bases of V if and only if there is a unitary matrix U such that $B = U^*AU = U^{-1}AU$. We describe this situation by saying that B is *unitarily similar* to A.

When the ground field is $\mathbb{R}$, the word *orthogonal* is often used instead of unitary. In this case A is orthogonal if and only if A^{-1} exists and is A^t. When there exists an orthogonal matrix U such that $B = U^tAU = U^{-1}AU$ then we say that B is *orthogonally similar* to A.

It is clear that the relation of being unitarily (or orthogonally) similar is an equivalence relation on the set of $n \times n$ matrices over $\mathbb{C}$ (or $\mathbb{R}$). Just as with ordinary similarity, the problem of locating particularly simple representatives, or *canonical forms*, in certain equivalence classes is important from both the theoretical and practical points of view. We shall consider this problem later.

CHAPTER EIGHT

Orthogonal direct sums

In 7.14 we obtained, in an inner product space V, a direct sum decomposition of the form $V = W \oplus W^\perp$. This leads us to consider the following notion.

Definition Let $V_1, \ldots, V_n$ be non-zero subspaces of an inner product space V. Then V is said to be the *orthogonal direct sum* of $V_1, \ldots, V_n$ if

(1) $V = \bigoplus_{i=1}^{n} V_i$;

(2) $(i = 1, \ldots, n)$ $\quad V_i^\perp = \sum_{j \neq i} V_j$.

In order to study orthogonal direct sum decompositions in an inner product space V let us begin by considering a projection $p : V \to V$ and the associated decomposition $V = \operatorname{Im} p \oplus \operatorname{Ker} p$ established in 2.6. In order that this be an orthogonal direct sum, it is clear that p has to be an *ortho-projection* in the sense that $\operatorname{Ker} p = (\operatorname{Im} p)^\perp$ or, equivalently, $\operatorname{Im} p = (\operatorname{Ker} p)^\perp$. To discover when this happens, we require the following result.

8.1 Theorem *If W, X are subspaces of a finite-dimensional inner product space V such that $V = W \oplus X$ then $V = W^\perp \oplus X^\perp$.*

Proof By 7.16 we have $\{0_V\} = V^\perp = (W + X)^\perp = W^\perp \cap X^\perp$ and $V = \{0_V\}^\perp = (W \cap X)^\perp = W^\perp + X^\perp$ and hence $V = W^\perp \oplus X^\perp$. $\diamond$

8.2 Corollary *If p is the projection on W parallel to X then $p^\star$ is the projection on $X^\perp$ parallel to $W^\perp$.*

Proof By 7.12 and since p is idempotent, we have $p^\star \circ p^\star = (p \circ p)^\star = p^\star$. Thus $p^\star$ is idempotent and so is the projection on $\operatorname{Im} p^\star$ parallel to $\operatorname{Ker} p^\star$. By 2.5, $\operatorname{Im} p = W$ and $\operatorname{Ker} p = X$ so, by 7.17, $W^\perp = (\operatorname{Im} p)^\perp = \operatorname{Ker} p^\star$ and $X^\perp = (\operatorname{Ker} p)^\perp = \operatorname{Im} p^\star$. $\diamond$

Definition If V is an inner product space then $f : V \to V$ is said to be *self-adjoint* if $f = f^\star$.

8.3 Theorem *Let V be an inner product space of finite dimension. If p is a projection on V then p is an ortho-projection if and only if p is self-adjoint.*

Proof By 8.2, $p^\star$ is the projection on $\operatorname{Im} p^\star = (\operatorname{Ker} p)^\perp$ parallel to $\operatorname{Ker} p^\star = (\operatorname{Im} p)^\perp$. If then p is an ortho-projection we have $\operatorname{Im} p^\star = \operatorname{Im} p$. It follows by 2.5 that for every $x \in V$ we have $p(x) = p^\star[p(x)]$. Consequently $p = p^\star \circ p$ and hence

$$p^\star = (p^\star \circ p)^\star = p^\star \circ p^{\star\star} = p^\star \circ p = p,$$

so that p is self-adjoint.

Conversely, if $p = p^\star$ then $\operatorname{Im} p = \operatorname{Im} p^\star = (\operatorname{Ker} p)^\perp$ shows that p is an ortho-projection. $\diamond$

It is clear from the above results that if V is an inner product space of finite dimension and if $V_1, \ldots, V_n$ are non-zero subspaces of V such that $V = \bigoplus_{i=1}^{n} V_i$ then this sum is an orthogonal direct sum if and only if, for every i, the projection p_i of V onto V_i parallel to $\sum_{j \neq i} V_j$ is self-adjoint.

It is also clear that if $V = \bigoplus_{i=1}^{n} V_i$ then this direct sum is an orthogonal direct sum if and only if, for each i, every element of V_i is orthogonal to every element of V_j when $j \neq i$. In fact in this case we have $\sum_{j \neq i} V_j \subseteq V_i^\perp$ whence we have equality since

$$\dim \sum_{j \neq i} V_j = \dim V - \dim V_i = \dim V_i^\perp.$$

Suppose that V is a finite-dimensional inner product space and that $f : V \to V$ is linear. We shall now consider under what conditions f is *ortho-diagonalizable* in the sense that there is an orthonormal basis of V consisting of eigenvectors of f; equivalently, under what conditions there is an ordered orthonormal basis of V with respect to which the matrix of f is diagonal. In purely matrix terms this problem is that of determining when a given square matrix (over $\mathbb{R}$ or $\mathbb{C}$) is unitarily similar to a diagonal matrix.

8.4 Theorem *Let V be a non-zero finite-dimensional inner product space over a field F and let $f : V \to V$ be linear. Then f is ortho-diagonalizable if and only if there are non-zero self-adjoint projections $p_1, \ldots, p_k : V \to V$ and distinct scalars $\lambda_1, \ldots, \lambda_k \in F$ such that*

(1) $f = \sum_{i=1}^{k} \lambda_i p_i$;

(2) $\sum_{i=1}^{k} p_i = \mathrm{id}_V$;

(3) $(i \neq j) \quad p_i \circ p_j = 0$.

Proof $\Rightarrow$: Since f is diagonalizable, we have $V = \bigoplus_{i=1}^{k} V_{\lambda_i}$ where $\lambda_1, \ldots, \lambda_k$ are the distinct eigenvalues of f and the subspace $V_{\lambda_i} = \mathrm{Ker}(f - \lambda_i \mathrm{id}_V)$ is the eigenspace associated with λ_i. If $p_i : V \to V$ is the projection on V_{λ_i} parallel to $\sum_{j \neq i} V_{\lambda_j}$ then (2), (3) follow from 2.8. Now for every $x \in V$ we have

$$f(x) = f\left(\sum_{i=1}^{k} p_i(x)\right) = \sum_{i=1}^{k} f[p_i(x)]$$
$$= \sum_{i=1}^{k} \lambda_i p_i(x) = \left(\sum_{i=1}^{k} \lambda_i p_i\right)(x)$$

and this gives (1). The fact that $\bigoplus_{i=1}^{k} V_{\lambda_i}$ is an orthogonal direct sum means that each projection p_i is an ortho-projection and so, by 8.3, is self-adjoint.

$\Leftarrow$: If the conditions hold then by 2.8 we have $V = \bigoplus_{i=1}^{k} \mathrm{Im}\, p_i$. Now the λ_i appearing in (1) are precisely the distinct eigenvalues of f. To see this, observe that

$$f \circ p_j = \left(\sum_{i=1}^{k} \lambda_i p_i\right) \circ p_j = \sum_{i=1}^{k} \lambda_i (p_i \circ p_j) = \lambda_j p_j$$

so $(f - \lambda_j \operatorname{id}_V) \circ p_j = 0$ and hence

$$\{0_V\} \neq \operatorname{Im} p_j \subseteq \operatorname{Ker}(f - \lambda_j \operatorname{id}_V).$$

Thus each λ_j is an eigenvalue of f. On the other hand, for every $\lambda \in F$ we have

$$f - \lambda \operatorname{id}_V = \sum_{i=1}^{k} \lambda_i p_i - \sum_{i=1}^{k} \lambda p_i = \sum_{i=1}^{k} (\lambda_i - \lambda) p_i$$

so that, if x is an eigenvector of f corresponding to the eigenvalue λ, $\sum_{i=1}^{k} (\lambda_i - \lambda) p_i(x) = 0_V$ and hence, since $V = \bigoplus_{i=1}^{k} \operatorname{Im} p_i$, we have $(\lambda_i - \lambda) p_i(x) = 0_V$ for $i = 1, \ldots, k$. If $\lambda \neq \lambda_i$ for every i then $p_i(x) = 0_V$ for every i and we have the contradiction $x = \sum_{i=1}^{k} p_i(x) = 0_V$. Thus $\lambda = \lambda_i$ for some i and consequently $\lambda_1, \ldots, \lambda_k$ are the distinct eigenvalues of f.

We now observe that $\operatorname{Im} p_j = \operatorname{Ker}(f - \lambda_j \operatorname{id}_V)$. For, suppose that $f(x) = \lambda_j x$. Then $0_V = \sum_{i=1}^{k} (\lambda_i - \lambda_j) p_i(x)$ and therefore $(\lambda_i - \lambda_j) p_i(x) = 0_V$ for all i whence $p_i(x) = 0_V$ for all $i \neq j$. Then $x = \sum_{i=1}^{k} p_i(x) = p_j(x) \in \operatorname{Im} p_j$ and so $\operatorname{Ker}(f - \lambda_j \operatorname{id}_V) \subseteq \operatorname{Im} p_j$. The reverse inclusion was established above.

Since now $V = \bigoplus_{i=1}^{k} \operatorname{Im} p_i = \bigoplus_{i=1}^{k} \operatorname{Ker}(f - \lambda_i \operatorname{id}_V)$ it follows that V has a basis consisting of eigenvectors of f and so f is diagonalizable. Now by hypothesis the projections p_i are self-adjoint so, for $j \neq i$,

$$\langle p_i(x) \,|\, p_j(x) \rangle = \langle p_i(x) \,|\, p_j^\star(x) \rangle = \langle p_j[p_i(x)] \,|\, x \rangle = \langle 0_V \,|\, x \rangle = 0.$$

It follows that the above eigenvector basis is orthogonal. By normalising each vector in this basis we obtain an orthonormal basis of eigenvectors. Hence f is ortho-diagonalizable. $\diamond$

Definition For an ortho-diagonalizable mapping f the equality $f = \sum_{i=1}^{k} \lambda_i p_i$ of 8.4 is called the *spectral resolution* of f.

Suppose now that $f : V \to V$ is ortho-diagonalizable. Applying the results of 7.12 to the conditions in 8.4 we obtain, with an obvious notation,

$$(1^*) \quad f^* = \sum_{i=1}^{k} \overline{\lambda_i} p_i^* = \sum_{i=1}^{k} \overline{\lambda_i} p_i; \quad (2^*) = (2), \quad (3^*) = (3).$$

We deduce by 8.4 that f^* is also ortho-diagonalizable and that (1^*) gives its spectral resolution (so that $\overline{\lambda_1}, \ldots, \overline{\lambda_k}$ are the distinct eigenvalues of f^*). A simple calculation now reveals that

$$f \circ f^* = \sum_{i=1}^{k} |\lambda_i|^2 p_i = f^* \circ f$$

from which we deduce that *ortho-diagonalizable mappings commute with their adjoints*. This observation leads to the following notion.

Definition If V is a finite-dimensional inner product space and $f : V \to V$ is linear then we say that f is *normal* if it commutes with its adjoint. Similarly, a square matrix A over the ground field of V is said to be *normal* if $AA^* = A^*A$.

Example It is readily seen that

$$A = \begin{bmatrix} 1 & i \\ 1 & 2+i \end{bmatrix}$$

is normal.

We have just seen that a necessary condition for a linear mapping f to be ortho-diagonalizable is that it be normal. It is quite remarkable that, when the ground field is $\mathbb{C}$, this condition is also sufficient. In order to establish this, we require the following properties of normal linear mappings.

8.5 Theorem *Let V be a non-zero finite-dimensional inner product space and let $f : V \to V$ be normal. Then*
(1) $(\forall x \in V) \quad \|f(x)\| = \|f^*(x)\|$;
(2) *if p is a polynomial with coefficients in the ground field of V then $p(f) : V \to V$ is also normal*;
(3) $\operatorname{Im} f \cap \operatorname{Ker} f = \{0_V\}$.

Proof (1) Since $f \circ f^* = f^* \circ f$ we have, for all $x \in V$,

$$\langle f(x) \mid f(x) \rangle = \langle x \mid f^*[f(x)] \rangle = \langle x \mid f[f^*(x)] \rangle = \langle f^*(x) \mid f^*(x) \rangle$$

from which (1) follows.

(2) If $p = a_0 + a_1 X + \cdots + a_n X^n$ then $p(f) = a_0 \operatorname{id}_V + a_1 f + \cdots + a_n f^n$ and, by 7.12, $[p(f)]^* = \overline{a_0} \operatorname{id}_V + \overline{a_1} f^* + \cdots + \overline{a_n}(f^*)^n$. Since f and f^* commute, it follows that so do $p(f)$ and $[p(f)]^*$. Thus $p(f)$ is normal.

(3) If $x \in \operatorname{Im} f \cap \operatorname{Ker} f$ then there exists $y \in V$ such that $x = f(y)$ and $f(x) = 0_V$. By (1) we have $f^*(x) = 0_V$ and so

$$0 = \langle f^*(x) \mid y \rangle = \langle x \mid f(y) \rangle = \langle x \mid x \rangle$$

whence $x = 0_V$. $\diamond$

8.6 Theorem *Let V be a non-zero finite-dimensional inner product space. If p is a projection on V then p is normal if and only if it is self-adjoint.*

Proof Clearly, if p is self-adjoint then p is normal. Suppose, conversely, that p is normal. By 8.5 we have $\|p(x)\| = \|p^*(x)\|$ and so $p(x) = 0_V$ if and only if $p^*(x) = 0_V$. Given $x \in V$, let $y = x - p(x)$. We have $p(y) = p(x) - p(x) = 0_V$ and so $0_V = p^*(y) = p^*(x) - p^*[p(x)]$. Thus $p^* = p^* \circ p$ and so

$$p = p^{**} = (p^* \circ p)^* = p^* \circ p^{**} = p^* \circ p = p^*$$

i.e. p is self-adjoint. $\diamond$

We can now solve the ortho-diagonalization problem for *complex* inner product spaces.

8.7 Theorem *Let V be a non-zero finite-dimensional complex inner product space. If $f : V \to V$ is linear then f is ortho-diagonalizable if and only if f is normal.*

Proof We have already seen that the condition is necessary. As for sufficiency, suppose that f is normal. To show that f is diagonalizable, it suffices to show that its minimum polynomial is a product of distinct linear factors. For this, we make use of the fact that $\mathbb{C}$ is algebraically closed, in the sense that every polynomial over $\mathbb{C}$ of degree at least 1 can be expressed as a

product of linear polynomials. Thus m_f is certainly a product of linear polynomials. Suppose, by way of obtaining a contradiction, that $\alpha \in \mathbb{C}$ is a multiple zero of m_f, so that we have $m_f = (X - \alpha)^2 g$ for some polynomial g. Then for every $x \in V$ we have
$$0_V = [m_f(f)](x) = [(f - \alpha \operatorname{id}_V)^2 \circ g(f)](x)$$
and consequently $[(f - \alpha \operatorname{id}_V) \circ g(f)](x)$ belongs to both the image and the kernel of $f - \alpha \operatorname{id}_V$. Since, by 8.5(2), $f - \alpha \operatorname{id}_V$ is normal we deduce from 8.5(3) that
$$(\forall x \in V) \qquad [(f - \alpha \operatorname{id}_V) \circ g(f)](x) = 0_V.$$
Consequently $(f - \alpha \operatorname{id}_V) \circ g(f)$ is the zero mapping on V, and this contradicts the fact that $(X - \alpha)^2 g$ is the minimum polynomial of f. Thus we see that f is diagonalizable.

To show that f is ortho-diagonalizable, it suffices to show that the corresponding projections p_i of 8.4 are ortho-projections, and by 8.3 it is enough to show that they are self-adjoint. Now since f is diagonalizable it is clear from the proof of 2.10 that for each i there is a polynomial t_i such that $t_i(f) = p_i$. By 8.5(2), each p_i is therefore normal and so, by 8.6, is self-adjoint. $\Diamond$

8.8 Corollary *If A is a square matrix over $\mathbb{C}$ then A is unitarily similar to a diagonal matrix if and only if A is normal.* $\Diamond$

It should be noted that in the proof of 8.7 we made use of the fact that the field $\mathbb{C}$ is algebraically closed. This is not true of $\mathbb{R}$ and so we might expect that the corresponding result fails in general for real inner product spaces (and real square matrices). This is indeed the case : there exist normal linear mappings on a real inner product space that are not diagonalizable. One way in which this can happen is when all the eigenvalues of the mapping in question are complex. For example, the rotation matrix
$$R_{2\pi/3} = \begin{bmatrix} -\frac{1}{2} & -\frac{\sqrt{3}}{2} \\ \frac{\sqrt{3}}{2} & -\frac{1}{2} \end{bmatrix}$$
is normal and its minimum polynomial is $X^2 + X + 1$ which has no zeros in $\mathbb{R}$. So, in order to obtain an analogue of 8.7 in the case where the ground field is $\mathbb{R}$, we are led to consider normal linear mappings whose eigenvalues are all real. These can be characterised as follows.

8.9 Theorem *Let V be a non-zero finite-dimensional complex inner product space. If $f : V \to V$ is linear then the following conditions are equivalent:*

(1) *f is normal and all its eigenvalues are real;*
(2) *f is self-adjoint.*

Proof (1) $\Rightarrow$ (2) : By 8.7, f is ortho-diagonalizable. Let $f = \sum_{i=1}^{k} \lambda_i p_i$ be its spectral resolution. We know that $f^\star$ is also normal with spectral resolution $f^\star = \sum_{i=1}^{k} \overline{\lambda_i} p_i$. Since each λ_i is real by hypothesis, it follows that $f^\star = f$.

(2) $\Rightarrow$ (1) : If $f^\star = f$ then clearly f is normal. If $f = \sum_{i=1}^{k} \lambda_i p_i$ and $f^\star = \sum_{i=1}^{k} \overline{\lambda_i} p_i$ are the spectral resolutions then we have $\sum_{i=1}^{k}(\lambda_i - \overline{\lambda_i})p_i = 0$ and so $\sum_{i=1}^{k}(\lambda_i - \overline{\lambda_i})p_i(x) = 0_V$ for every $x \in V$, whence $(\lambda_i - \overline{\lambda_i})p_i = 0$ for every i since $V = \bigoplus_{i=1}^{k} \operatorname{Im} p_i$. Since no p_i can be zero, we deduce that $\lambda_i = \overline{\lambda_i}$ for every i and so every eigenvalue of f is real. $\diamond$

8.10 Corollary *All the eigenvalues of a self-adjoint matrix are real.* $\diamond$

The analogue of 8.7 is now the following.

8.11 Theorem *Let V be a non-zero finite-dimensional real inner product space. If $f : V \to V$ is linear then f is ortho-diagonalizable if and only if f is self-adjoint.*

Proof $\Rightarrow$: If f is ortho-diagonalizable let $f = \sum_{i=1}^{k} \lambda_i p_i$ be its spectral resolution. Since the ground field is $\mathbb{R}$, every λ_i is real and so, taking adjoints and using 8.3, we obtain $f^\star = f$.

$\Leftarrow$: Suppose, conversely, that f is self-adjoint and let A be the $(n \times n$, say) matrix of f relative to some ordered orthonormal basis of V. Then clearly A is symmetric. Now let f' be the linear mapping on the complex inner product space $\mathbb{C}^n$ whose matrix, relative to the standard ordered (orthonormal) basis, is

ORTHOGONAL DIRECT SUMS   95

A. Then f' is self-adjoint. By 8.9, the eigenvalues of f' are all real and, since f' is diagonalizable, the minimum polynomial of f' is a product of distinct linear polynomials over $\mathbb{R}$. Since this is then the minimum polynomial of A, it is also the minimum polynomial of f. Thus we see that f is diagonalizable. That it is ortho-diagonalizable is established precisely as in 8.7. ◇

8.12 Corollary *If A is a square matrix over $\mathbb{R}$ then A is orthogonally similar to a diagonal matrix if and only if A is symmetric.* ◇

Example By 8.12, the real symmetric matrix

$$A = \begin{bmatrix} 4 & 2 & 2 \\ 2 & 4 & 2 \\ 2 & 2 & 4 \end{bmatrix}$$

is orthogonally similar to a diagonal matrix D. Let us find an orthogonal matrix P such that $P^{-1}AP = D$. It is readily seen that $\chi_A = (X-2)^2(X-8)$. A basis for the eigenspace associated with the eigenvalue 2 is

$$\{[-1\ 1\ 0]^t, [-1\ 0\ 1]^t\}.$$

Applying the Gram-Schmidt process, we see that an orthonormal basis for this eigenspace is

$$\{[-\tfrac{1}{\sqrt{2}}\ \tfrac{1}{\sqrt{2}}\ 0]^t, [-\tfrac{1}{\sqrt{6}}\ -\tfrac{1}{\sqrt{6}}\ \tfrac{2}{\sqrt{6}}]^t\}.$$

As for the eigenspace associated with the eigenvalue 8, a basis is $\{[1\ 1\ 1]^t\}$ so an orthonormal basis is $\{[\tfrac{1}{\sqrt{3}}\ \tfrac{1}{\sqrt{3}}\ \tfrac{1}{\sqrt{3}}]^t\}$. Pasting these orthonormal bases together, we see that the matrix

$$P = \begin{bmatrix} -1/\sqrt{2} & -1/\sqrt{6} & 1/\sqrt{3} \\ 1/\sqrt{2} & -1/\sqrt{6} & 1/\sqrt{3} \\ 0 & 2/\sqrt{6} & 1/\sqrt{3} \end{bmatrix}$$

is orthogonal and such that $P^{-1}AP = \text{diag}\{2\ 2\ 8\}$.

We shall now derive a useful alternative characterisation of a self-adjoint linear mapping on a complex inner product space. In order to do so, we require the following result.

8.13 Theorem *Let V be a complex inner product space. If $f : V \to V$ is linear and such that $\langle f(x) \,|\, x \rangle = 0$ for all $x \in V$ then $f = 0$.*

Proof Using the condition we see that, for all $z \in V$,
$$0 = \langle f(y+z) \,|\, y+z \rangle = \langle f(y) \,|\, z \rangle + \langle f(z) \,|\, y \rangle;$$
$$0 = \langle f(iy+z) \,|\, iy+z \rangle = i\langle f(y) \,|\, z \rangle - i\langle f(z) \,|\, y \rangle,$$
from which it follows that $\langle f(y) \,|\, z \rangle = 0$. Then $f(y) = 0_V$ for all $y \in V$ and so $f = 0$. $\diamond$

8.14 Theorem *Let V be a finite-dimensional complex inner product space and let $f : V \to V$ be linear. Then f is self-adjoint if and only if $\langle f(x) \,|\, x \rangle \in \mathbb{R}$ for all $x \in V$.*

Proof $\Rightarrow$: If f is self-adjoint then for every $x \in V$ we have
$$\overline{\langle f(x) \,|\, x \rangle} = \overline{\langle f^\star(x) \,|\, x \rangle} = \langle x \,|\, f^\star(x) \rangle = \langle f(x) \,|\, x \rangle,$$
from which the result follows.

$\Leftarrow$: If $\langle f(x) \,|\, x \rangle \in \mathbb{R}$ for every $x \in V$ then
$$\langle f^\star(x) \,|\, x \rangle = \langle x \,|\, f(x) \rangle = \overline{\langle f(x) \,|\, x \rangle} = \langle f(x) \,|\, x \rangle$$
and consequently
$$\langle (f^\star - f)(x) \,|\, x \rangle = \langle f^\star(x) \,|\, x \rangle - \langle f(x) \,|\, x \rangle = 0.$$
Since this holds for all $x \in V$ it follows by 8.13 that $f^\star = f$. $\diamond$

This now leads to the following notion.

Definition If V is an inner product space then a linear mapping $f : V \to V$ is said to be *positive* (or *semi-definite*) if it is self-adjoint and such that $\langle f(x) \,|\, x \rangle \geq 0$ for every $x \in V$; and *positive definite* if it is self-adjoint and such that $\langle f(x) \,|\, x \rangle > 0$ for every non-zero $x \in V$.

8.15 Theorem *If V is a non-zero finite-dimensional inner product space and $f : V \to V$ is linear then the following statements are equivalent :*

(1) *f is positive;*
(2) *f is self-adjoint and every eigenvalue is real and greater than or equal to 0;*
(3) *there is a self-adjoint $g : V \to V$ such that $g^2 = f$;*
(4) *there exists $h : V \to V$ such that $h^\star \circ h = f$.*

Proof (1) ⇒ (2) : Let λ be an eigenvalue of f. By 8.9 (in the complex case), λ is real. Then $0 \le \langle f(x) | x \rangle = \langle \lambda x | x \rangle = \lambda \langle x | x \rangle$ gives $\lambda \ge 0$ since $\langle x | x \rangle > 0$.

(2) ⇒ (3) : Since f is self-adjoint it is normal and hence is ortho-diagonalizable. Let its spectral resolution be $f = \sum_{i=1}^{k} \lambda_i p_i$ and, using (2), define $g : V \to V$ by $g = \sum_{i=1}^{k} \sqrt{\lambda_i} p_i$. Since the p_i are ortho-projections, and hence self-adjoint, we have that g is self-adjoint. Also, since $p_i \circ p_j = 0$ when $i \ne j$, it follows readily that $g^2 = f$.

(3) ⇒ (4) : Take $h = g$.

(4) ⇒ (1) : Observe that $(h^* \circ h)^* = h^* \circ h^{**} = h^* \circ h$ and, for all $x \in V$, $\langle h^*[h(x)] | x \rangle = \langle h(x) | h(x) \rangle \ge 0$. Thus we see that $h^* \circ h$ is positive. ◇

It is immediate from 8.15 that every positive linear mapping f has a square root. We shall now show in fact that f has a unique positive square root.

8.16 Theorem *Let f be a positive linear mapping on a non-zero finite-dimensional inner product space V. Then there is a unique positive linear mapping $g : V \to V$ such that $g^2 = f$. Moreover, there is a polynomial q such that $g = q(f)$.*

Proof Let $f = \sum_{i=1}^{k} \lambda_i p_i$ be the spectral resolution of f and define g as before, namely $g = \sum_{i=1}^{k} \sqrt{\lambda_i} p_i$. Since this must be the spectral resolution of g, it follows that the eigenvalues of g are $\sqrt{\lambda_i}$ for $i = 1, \ldots, k$ and so, by 8.15, g is positive.

Suppose now that $h : V \to V$ is also positive and such that $h^2 = f$. If the spectral resolution of h is $\sum_{j=1}^{m} \mu_j q_j$ where the q_j are orthogonal projections then we have

$$\sum_{i=1}^{k} \lambda_i p_i = f = h^2 = \sum_{j=1}^{m} \mu_j^2 q_j.$$

As in the proof of 8.4, the eigenspaces of f are $\text{Im } p_i$ for $i = 1, \ldots, k$ and also $\text{Im } q_j$ for $j = 1, \ldots, m$. It follows that $m = k$

and that there is a permutation σ on $1,\ldots,k$ such that $q_{\sigma(i)} = p_i$, whence $\mu_{\sigma(i)}^2 = \lambda_i$. Thus $\mu_{\sigma(i)} = \sqrt{\lambda_i}$ and we deduce that $h = g$.

The final statement follows by considering the Lagrange polynomials
$$P_i = \prod_{j \neq i} \frac{X - \lambda_j}{\lambda_i - \lambda_j}.$$
Since $P_i(\lambda_t) = \delta_{it}$, the polynomial $q = \sum_{i=1}^{k} \sqrt{\lambda_i} P_i$ is then such that $q(f) = g$. $\diamond$

There is a corresponding result to 8.15 for positive definite linear mappings, namely :

8.17 Theorem *If V is a non-zero finite-dimensional inner product space and $f : V \to V$ is linear then the following statements are equivalent :*

(1) *f is positive definite;*
(2) *f is self-adjoint and all eigenvalues of f are real and strictly positive;*
(3) *there is an invertible self-adjoint g such that $g^2 = f$;*
(4) *there is an invertible h such that $h^* \circ h = f$.*

Proof This follows immediately from 8.15 on noting that g is invertible if and only if 0 is not one of its eigenvalues. $\diamond$

8.18 Corollary *If f is positive definite then f is invertible.* $\diamond$

Of course 8.15 to 8.18 have matrix analogues. A square matrix that represents a positive linear mapping with respect to some ordered orthonormal basis in a finite-dimensional inner product space is often called a *Gram matrix*. By 8.15 we have the following characterisation.

8.19 Theorem *A square matrix is a Gram matrix if and only if it is self-adjoint and all its eigenvalues are greater than or equal to 0.* $\diamond$

CHAPTER NINE

Bilinear and quadratic forms

In this Chapter we shall apply some of the previous results in a study of certain types of linear forms.

Definition Let V be a finite-dimensional vector space over a field F. A *bilinear form* on V is a mapping $f : V \times V \to F$ such that the following identities hold :
(1) $f(x + x', y) = f(x, y) + f(x', y)$;
(2) $f(x, y + y') = f(x, y) + f(x, y')$;
(3) $f(\lambda x, y) = \lambda f(x, y) = f(x, \lambda y)$.

Example The 'dot product' on $\mathbb{R}^n$, namely that given by
$$f\big((x_1, \ldots, x_n), (y_1, \ldots, y_n)\big) = \sum_{i=1}^{n} x_i y_i$$
is a bilinear form on $\mathbb{R}^n$. This is readily seen if we consider the right hand side as the product of a row matrix and a column matrix (and commit the usual abuse in identifying a scalar with a 1×1 matrix).

If now V is of dimension n over F let $(v_i)_n$ be an ordered basis of V. If $f : V \times V \to F$ is a bilinear form on V then by the *matrix of f relative to $(v_i)_n$* we shall mean the $n \times n$ matrix $A = [a_{ij}]$ given by
$$a_{ij} = f(v_i, v_j).$$
Note that if $x = \sum_{i=1}^{n} x_i v_i$ and $y = \sum_{i=1}^{n} y_i v_i$ then, by the bilinearity of f, we have
$$(\star) \qquad f(x, y) = \sum_{i=1}^{n} \sum_{j=1}^{n} x_i y_j f(v_i, v_j) = \sum_{i,j=1}^{n} x_i y_j a_{ij}.$$

Conversely, given any $n \times n$ matrix $A = [a_{ij}]$ over F it is easy to see that $(\star)$ defines a bilinear form on V; simply observe that the right hand side can be written as

$$\mathbf{x}^t A \mathbf{y} = [x_1 \ \ldots \ x_n] A \begin{bmatrix} y_1 \\ \vdots \\ y_n \end{bmatrix}.$$

Moreover, with respect to the ordered basis $(v_i)_n$ the matrix of f is A.

Example In the previous Example the matrix of f relative to the standard ordered basis of $\mathbb{R}^n$ is I_n.

Example The matrix

$$A = \begin{bmatrix} a & h & g \\ h & b & f \\ g & f & c \end{bmatrix}$$

gives rise to the bilinear form $\vartheta : \mathbb{R}^3 \times \mathbb{R}^3 \to \mathbb{R}$ described by

$$\mathbf{x}^t A \mathbf{y} = ax_1y_1 + bx_2y_2 + cx_3y_3 + h(x_1y_2 + x_2y_1)$$
$$+ g(x_1y_3 + x_3y_1) + f(x_2y_3 + x_3y_2).$$

Example The matrix

$$A = \begin{bmatrix} 0 & 1 & 1 \\ 0 & 0 & 1 \\ 0 & 0 & 0 \end{bmatrix}$$

gives rise to the bilinear form $\vartheta : \mathbb{R}^3 \times \mathbb{R}^3 \to \mathbb{R}$ described by

$$\mathbf{x}^t A \mathbf{y} = x_1(y_2 + y_3) + x_2 y_3.$$

It is natural to ask how the matrix of a bilinear form changes when we change reference to another ordered basis.

9.1 Theorem *Let V be a vector space of dimension n over a field F. Let $(v_i)_n$ and $(w_i)_n$ be ordered bases of V. If $f : V \times V \to F$ is bilinear and if A is the matrix of f relative to $(v_i)_n$ then the matrix of f relative to $(w_i)_n$ is $P^t A P$ where P is the transition matrix from $(v_i)_n$ to $(w_i)_n$.*

Proof We have $w_j = \sum_{i=1}^{n} p_{ij} v_i$ for $j = 1, \ldots, n$ and so, by the bilinearity of f,

$$\begin{aligned} f(w_i, w_j) &= f\left(\sum_{t=1}^{n} p_{ti} v_t, \sum_{k=1}^{n} p_{kj} v_k\right) \\ &= \sum_{t=1}^{n} \sum_{k=1}^{n} p_{ti} p_{kj} f(v_t, v_k) \\ &= \sum_{t=1}^{n} \sum_{k=1}^{n} p_{ti} p_{kj} a_{tk} \\ &= [P^t A P]_{ij}, \end{aligned}$$

from which the result follows. $\diamondsuit$

Definition If A, B are $n \times n$ matrices over a field F then we say that B is *congruent* to A if there is an invertible matrix P such that $B = P^t A P$.

It is readily seen that the relation of being congruent is an equivalence relation on $\text{Mat}_{n \times n}(F)$.

Definition A bilinear form $f : V \times V \to F$ is said to be *symmetric* if

$$(\forall x, y \in V) \quad f(x, y) = f(y, x).$$

It is clear that a matrix that represents a symmetric bilinear form is symmetric; and, conversely, that every symmetric matrix gives rise to a symmetric bilinear form.

Definition If V is a vector space over a field F then by a *quadratic form* on V we mean a mapping $Q : V \to F$ such that, for some symmetric bilinear form $f : V \times V \to F$, we have $Q(x) = f(x, x)$ for all $x \in V$.

Example $Q : \mathbb{R}^2 \to \mathbb{R}$ given by $Q(x, y) = x^2 - xy + y^2$ is a quadratic form. This is clear from the fact that we can write the right hand side as

$$\mathbf{x}^t A \mathbf{x} = \begin{bmatrix} x & y \end{bmatrix} \begin{bmatrix} 1 & -\frac{1}{2} \\ -\frac{1}{2} & 1 \end{bmatrix} \begin{bmatrix} x \\ y \end{bmatrix}.$$

Example $Q : \mathbb{R}^3 \to \mathbb{R}$ given by $Q(x,y,z) = x^2 + y^2 - z^2$ is a quadratic form, for we can write the right hand side as

$$\mathbf{x}^t A \mathbf{x} = \begin{bmatrix} x & y & z \end{bmatrix} \begin{bmatrix} 1 & & \\ & 1 & \\ & & -1 \end{bmatrix} \begin{bmatrix} x \\ y \\ z \end{bmatrix}.$$

In what follows we shall restrict the ground field F to be $\mathbb{R}$. Our reason for so doing is that most applications involve $\mathbb{R}$, and that there are certain difficulties that have to be avoided if F is such that $1_F + 1_F = 0_F$, i.e. when F is of characteristic 2.

Given a symmetric bilinear form $f : V \times V \to \mathbb{R}$ we shall denote by $Q_f : V \to \mathbb{R}$ the quadratic form given by $Q_f(x) = f(x,x)$.

9.2 Theorem *Let V be a vector space over $\mathbb{R}$. If $f : V \times V \to \mathbb{R}$ is a symmetric bilinear form then the following identities hold :*

(1) $Q_f(\lambda x) = \lambda^2 Q_f(x)$;
(2) $f(x,y) = \frac{1}{2}[Q_f(x+y) - Q_f(x) - Q_f(y)]$;
(3) $f(x,y) = \frac{1}{4}[Q_f(x+y) - Q_f(x-y)]$.

Proof (1) $Q_f(\lambda x) = f(\lambda x, \lambda x) = \lambda^2 f(x,x) = \lambda^2 Q_f(x)$.

(2) Since f is symmetric, we have

$$Q_f(x+y) = f(x+y, x+y) = Q_f(x) + 2f(x,y) + Q_f(y).$$

(3) By (1) we have $Q_f(-x) = Q_f(x)$ and so, by (2),

$$Q_f(x-y) = Q_f(x) - 2f(x,y) + Q_f(y).$$

This, together with (2), gives (3). $\diamond$

9.3 Corollary *Every real quadratic form is associated with a uniquely determined symmetric bilinear form.*

Proof If $f, g : V \times V \to \mathbb{R}$ are symmetric bilinear forms and $Q : V \to \mathbb{R}$ is a quadratic form such that $Q = Q_f = Q_g$ then by 9.2(2) it follows that $f = g$. $\diamond$

By the *matrix of a real quadratic form* we shall mean, by abuse of language, the matrix of the associated symmetric bilinear form.

Example $Q : \mathbb{R}^2 \to \mathbb{R}$ given by $Q(x,y) = 4x^2 + 6xy + 9y^2$ is a quadratic form. Its matrix is

$$\begin{bmatrix} 4 & 3 \\ 3 & 9 \end{bmatrix}$$

and the associated symmetric bilinear form is

$$f\big((x,y),(x',y')\big) = 4xx' + 3(xy' + x'y) + 9yy'.$$

It is clear that symmetric matrices A, B represent the same quadratic form relative to different ordered bases if and only if they are congruent. Our objective now is to obtain a canonical form for real symmetric matrices under congruence, i.e. a particularly simple representative in each congruence class. This will then give us a 'canonical form' for the associated real quadratic form. The results on orthogonal similarity that we have obtained in the previous Chapter will put us well on the road.

9.4 Theorem *If A is a real symmetric matrix then A is congruent to a unique matrix of the form*

$$\begin{bmatrix} I_r & & \\ & -I_s & \\ & & 0 \end{bmatrix}.$$

Proof Since A is real symmetric it follows by 8.12 and 8.10 that A is orthogonally similar to a diagonal matrix and all the eigenvalues of A are real. Let the positive eigenvalues be $\lambda_1, \ldots, \lambda_r$ and let the negative eigenvalues be $-\lambda_{r+1}, \ldots, -\lambda_{r+s}$. Then there is an orthogonal matrix P such that

$$P^t A P = \text{diag}\{\lambda_1, \ldots, \lambda_r, -\lambda_{r+1}, \ldots, -\lambda_{r+s}, 0, \ldots, 0\}.$$

Now let N be the diagonal matrix whose entries are

$$n_{ii} = \begin{cases} \frac{1}{\sqrt{\lambda_i}} & \text{if } i = 1, \ldots, r+s; \\ 1 & \text{otherwise.} \end{cases}$$

Then it is readily seen that

$$(PN)^t A P N = \begin{bmatrix} I_r & & \\ & -I_s & \\ & & 0 \end{bmatrix}.$$

Since P and N are each invertible, so is PN. Thus A is congruent to a matrix of the stated form.

As for uniqueness, it suffices to suppose that

$$L = \begin{bmatrix} I_r & & \\ & -I_s & \\ & & 0 \end{bmatrix}, \qquad M = \begin{bmatrix} I_{r'} & & \\ & -I_{s'} & \\ & & 0 \end{bmatrix}$$

are congruent and show that $r = r'$ and $s = s'$. Now if L and M are congruent then it is clear that they have the same rank and so $r + s = r' + s'$. Suppose, by way of obtaining a contradiction, that $r < r'$ (in which case $s' < s$). Let W be the real vector space $\text{Mat}_{n \times 1}(\mathbb{R})$. Clearly, W is an inner product space under the definition $\langle \mathbf{x} \,|\, \mathbf{y} \rangle = \mathbf{x}^t \mathbf{y}$. Consider the mapping $f_L : W \to W$ given by $f_L(\mathbf{x}) = L\mathbf{x}$. Since L is symmetric we have $\langle L\mathbf{x} \,|\, \mathbf{y} \rangle = \mathbf{x}^t L \mathbf{y} = \langle \mathbf{x} \,|\, L\mathbf{y} \rangle$ and so f_L is self-adjoint. Similarly, so is $f_M : W \to W$ where $f_M(\mathbf{x}) = M\mathbf{x}$. Consider now the subspaces

$$X = \{\mathbf{x} \in W \,;\, x_1 = \cdots = x_r = 0, x_{r+s+1} = \cdots = x_n = 0\};$$
$$Y = \{\mathbf{x} \in W \,;\, x_{r'+1} = \cdots = x_{r'+s'} = 0\}.$$

Clearly, X is of dimension s and for every non-zero $\mathbf{x} \in X$ we have

(1) $$\mathbf{x}^t L \mathbf{x} = -x_{r+1}^2 - \cdots - x_{r+s}^2 < 0.$$

Also, Y is of dimension $n - s'$ and for every $\mathbf{x} \in Y$ we have

$$\mathbf{x}^t M \mathbf{x} = x_1^2 + \cdots + x_r^2 \geq 0.$$

Now since L and M are congruent there is an invertible matrix P such that $M = P^t L P$. For all $\mathbf{x} \in Y$ we then have

$$0 \leq \mathbf{x}^t M \mathbf{x} = \langle M\mathbf{x} \,|\, \mathbf{x} \rangle = \langle P^t L P \mathbf{x} \,|\, \mathbf{x} \rangle$$
$$= \langle (f_L \circ f_P)(\mathbf{x}) \,|\, f_P(\mathbf{x}) \rangle,$$

from which we see that if $Z = \{f_P(\mathbf{x}) \,;\, \mathbf{x} \in Y\}$ then

(2) $$(\forall \mathbf{z} \in Z) \qquad \langle f_L(\mathbf{z}) \,|\, \mathbf{z} \rangle \geq 0.$$

Now since f_P is an isomorphism we have $\dim Z = \dim Y = n - s'$ and so

$$\dim Z + \dim X = n - s' + s > n \geq \dim(Z + X).$$

It follows that the sum $Z + X$ is not direct (otherwise we would have equality) and so $Z \cap X \neq \{0_V\}$. Let $\mathbf{z}$ be a non-zero element of $Z \cap X$. Then from (1) we see that $\langle f_L(\mathbf{z}) | \mathbf{z} \rangle$ is negative, whereas from (2) we see that $\langle f_L(\mathbf{z}) | \mathbf{z} \rangle$ is non-negative. This contradiction shows that we cannot have $r' < r$. Similarly we cannot have $r < r'$ and so we conclude that $r = r'$ whence also $s = s'$. $\diamondsuit$

The above result gives immediately the following theorem which describes canonical quadratic forms.

9.5 Theorem [Sylvester] *Let V be a vector space of dimension n over $\mathbb{R}$ and let $Q : V \to \mathbb{R}$ be a quadratic form on V. Then there is an ordered basis $(v_i)_n$ of V such that if $x = \sum_{i=1}^{n} x_i v_i$ then*

$$Q(x) = x_1^2 + \cdots + x_r^2 - x_{r+1}^2 - \cdots - x_{r+s}^2.$$

Moreover, the integers r and s are independent of such a basis. $\diamondsuit$

The integer $r + s$ in 9.5 is often called the *rank* of the quadratic form Q, and $r - s$ the *signature* of Q.

Example Consider the quadratic form $Q : \mathbb{R}^3 \to \mathbb{R}$ given by

$$Q(x, y, z) = x^2 - 2xy + 4yz - 2y^2 + 4z^2.$$

By the process of 'completing the squares' it is readily seen that

$$Q(x, y, z) = (x - y)^2 - 4y^2 + (y + 2z)^2$$

which is in canonical form, of rank 3 and signature 1. Alternatively, we can use matrices. The matrix of Q is

$$A = \begin{bmatrix} 1 & -1 & 0 \\ -1 & -2 & 2 \\ 0 & 2 & 4 \end{bmatrix}.$$

Let P be an orthogonal matrix such that $P^t A P$ is the diagonal matrix D. If $\mathbf{y} = P^t \mathbf{x}$ (so that $\mathbf{x} = P\mathbf{y}$) then

$$\mathbf{x}^t A \mathbf{x} = (P\mathbf{y})^t A P \mathbf{y} = \mathbf{y}^t P^t A P \mathbf{y} = \mathbf{y}^t D \mathbf{y},$$

where the right hand side is of the form $X^2 - 4Y^2 + Z^2$.

Example The quadratic form given by $Q(x, y, z) = 2xy + 2yz$ can be reduced to canonical form either by the method of completing squares or by a matrix reduction. The former method is not so easy in this case, but can be achieved as follows. Define

$$\sqrt{2}x = X + Y, \quad \sqrt{2}y = X - Y, \quad \sqrt{2}z = Z.$$

Then the form becomes

$$\begin{aligned} X^2 - Y^2 + (X - Y)Z &= (X + \tfrac{1}{2}Z)^2 - (Y + \tfrac{1}{2}Z)^2 \\ &= \tfrac{1}{2}(x + y + z)^2 - \tfrac{1}{2}(x - y + z)^2, \end{aligned}$$

which is of rank 2 and signature 0.

Definition A quadratic form Q is said to be *positive definite* if $Q(x) > 0$ for all non-zero x.

By taking the inner product space V to be $\text{Mat}_{n \times 1}(\mathbb{R})$ under $\langle \mathbf{x} \,|\, \mathbf{y} \rangle = \mathbf{x}^t \mathbf{y}$, we see that a quadratic form Q on V is positive definite if and only if, for all non-zero $\mathbf{x} \in V$,

$$0 < Q(\mathbf{x}) = \mathbf{x}^t A \mathbf{x} = \langle A\mathbf{x} \,|\, \mathbf{x} \rangle,$$

which is the case if and only if A is positive definite. It is clear that this situation obtains when there are no negative terms in the canonical form, i.e. when the rank and the signature are the same.

Example Let $f : \mathbb{R} \times \mathbb{R} \to \mathbb{R}$ be a function whose partial derivatives f_x, f_y are zero at (x_0, y_0). Then the Taylor series at $(x_0 + h, y_0 + k)$ is

$$f(x_0, y_0) + \tfrac{1}{2}[h^2 f_{xx} + 2hk f_{xy} + k^2 f_{yy}](x_0, y_0) + \cdots.$$

For small values of h, k the significant term is this quadratic form in h, k. If it has rank 2 then its normal form is $\pm H^2 \pm K^2$. If both signs are positive (i.e. the form is positive definite) then f has a relative minimum at (x_0, y_0), and if both signs are negative then f has a relative maximum at (x_0, y_0). If one sign is positive and the other is negative then f has a saddle-point at (x_0, y_0). Thus the geometry is distinguished by the signature of the quadratic form.

Example Consider the quadratic form

$$4x^2 + 4y^2 + 4z^2 - 2xy - 2yz + 2xz.$$

Its matrix is

$$A = \begin{bmatrix} 4 & -1 & 1 \\ -1 & 4 & -1 \\ 1 & -1 & 4 \end{bmatrix}.$$

The eigenvalues of A are 3 (of algebraic multiplicity 2), and 6. If P is an orthogonal matrix such that $P^t A P$ is diagonal then, changing coordinates by $\mathbf{X} = P^t \mathbf{x}$, we transform the quadratic form to

$$3X^2 + 3Y^2 + 6Z^2$$

which is positive definite.

CHAPTER TEN

Real normality

We have seen in 8.7 that the ortho-diagonalizable linear mappings on a complex inner product space are precisely those that are normal; and in 8.11 that the ortho-diagonalizable linear mappings on a real inner product space are precisely those that are self-adjoint. It is therefore natural to ask what can be said about *normal* linear mappings on a *real* inner product space; equivalently, to ask about real square matrices that commute with their transposes. Our objective now will be to obtain a canonical form for such a matrix under orthogonal similarity. For this purpose, we consider the following notion.

Definition Let V be a finite-dimensional real inner product space and let $f : V \to V$ be linear. We say that f is *skew-adjoint* if $f^\star = -f$. The corresponding terminology for real square matrices is *skew-symmetric*.

10.1 Theorem *If V is a non-zero finite-dimensional real inner product space and $f : V \to V$ is linear then there is a unique self-adjoint $g : V \to V$ and a unique skew-adjoint $h : V \to V$ such that $f = g + h$. Moreover, f is normal if and only if g, h commute.*

Proof We have $f = \frac{1}{2}(f + f^\star) + \frac{1}{2}(f - f^\star)$ where $\frac{1}{2}(f + f^\star)$ is self-adjoint and $\frac{1}{2}(f - f^\star)$ is skew-adjoint. Also, if $f = g + h$ where g is self-adjoint and h is skew-adjoint then $f^\star = g^\star + h^\star = g - h$ and consequently we see that $g = \frac{1}{2}(f + f^\star)$ and $h = \frac{1}{2}(f - f^\star)$.

Now $f \circ f^\star = f^\star \circ f$ gives $(g + h) \circ (g - h) = (g - h) \circ (g + h)$ which reduces to $g \circ h = h \circ g$. Conversely, if g, h commute then it is readily seen that $f \circ f^\star = g^2 - h^2 = f^\star \circ f$. $\diamond$

We now obtain a useful characterisation of skew-adjoint mappings.

10.2 Theorem *If V is a non-zero finite-dimensional real inner product space then $f : V \to V$ is skew-adjoint if and only if*

$$(\forall x \in V) \qquad \langle f(x) \,|\, x \rangle = 0.$$

Proof $\Rightarrow$: If f is skew-adjoint then for every $x \in V$ we have

$$\langle f(x) \,|\, x \rangle = \langle x \,|\, -f(x) \rangle = -\langle x \,|\, f(x) \rangle = -\langle f(x) \,|\, x \rangle$$

and so $\langle f(x) \,|\, x \rangle = 0$.

$\Leftarrow$: If the condition holds then for all $x, y \in V$ we have

$$0 = \langle f(x+y) \,|\, x+y \rangle = \langle f(x) \,|\, y \rangle + \langle f(y) \,|\, x \rangle$$

which gives

$$\langle f(x) \,|\, y \rangle = -\langle f(y) \,|\, x \rangle = -\langle x \,|\, f(y) \rangle = \langle x \,|\, -f(y) \rangle.$$

It now follows by the uniqueness of adjoints that $f^* = -f$. $\diamond$

As we shall see, the main results that we shall obtain will stem from applications of the Primary Decomposition Theorem. The notion of minimum polynomial will therefore play an important part in this. Now as the ground field is $\mathbb{R}$ we need to know what the monic irreducible polynomials over $\mathbb{R}$ look like. We recall (see, for example, Volume Three) that *a monic polynomial of degree at least one over $\mathbb{R}$ is irreducible if and only if it is of the form $X - a$, or $X^2 - 2aX + a^2 + b^2$ with $b \neq 0$.*

We begin with the following observation.

10.3 Theorem *If V is a non-zero finite-dimensional real inner product space and $f : V \to V$ is normal then $m_f = \prod_{i=1}^{k} p_i$ where $p_1, \ldots, p_k$ are distinct irreducible polynomials.*

Proof We know that the minimum polynomial of f is of the general form $m_f = \prod_{i=1}^{k} p_i^{e_i}$ where $p_1, \ldots, p_k$ are distinct irreducibles. What we have to show is that when f is normal every $e_i = 1$. Suppose, by way of obtaining a contradiction, that

$e_i \geq 2$ for a particular i. If $V_i = \operatorname{Ker} p_i(f)^{e_i}$ then for every $x \in V_i$ we have

$$p_i(f)^{e_i-1}(x) \in \operatorname{Im} p_i(f) \cap \operatorname{Ker} p_i(f).$$

But f is normal, hence so is $p_i(f)$ by 8.5(2). It now follows by 8.5(3) that the restriction of $p_i(f)^{e_i-1}$ to V_i is the zero mapping. If $f_i : V_i \to V_i$ is the mapping induced by f on the f-invariant subspace V_i we thus have $p_i(f_i)^{e_i-1} = 0$, and this contradicts the fact that, by 2.10, the minimum polynomial of f_i is $p_i^{e_i}$. $\diamond$

Concerning the minimum polynomial of a skew-adjoint mapping, we now have the following result.

10.4 Theorem *Let V be a non-zero finite-dimensional real inner product space and let $f : V \to V$ be a skew-adjoint linear mapping. If p is an irreducible factor of the minimum polynomial of f then either $p = X$ or $p = X^2 + c^2$ for some $c \neq 0$.*

Proof Since skew-adjoint mappings are normal it follows by 10.3 that m_f is of the form $\prod_{i=1}^{k} p_i$ where $p_1, \ldots, p_k$ are distinct irreducible polynomials. We also know that each p_i is either linear or of the form $p_i = X^2 - 2a_i X + a_i^2 + b_i^2$ where $b_i \neq 0$.

If p_i is linear, say $p_i = X - a_i$, and if f_i is the mapping induced on the primary component $V_i = \operatorname{Ker} p_i(f)$, then $f_i = a_i \operatorname{id}_{V_i}$ and consequently $f_i^* = f_i$. Since f_i is also skew-adjoint, it follows that $0 = f_i = a_i \operatorname{id}_{V_i}$. Thus $a_i = 0$ and $p_i = X$.

If now p_i is not linear then we have

$$0 = p_i(f_i) = f_i^2 - 2a_i f_i + (a_i^2 + b_i^2) \operatorname{id}_{V_i}.$$

Since f_i is skew-adjoint we deduce that

$$0 = f_i^2 + 2a_i f_i + (a_i^2 + b_i^2) \operatorname{id}_{V_i}.$$

These equalities give $4a_i f_i = 0$. Now $f_i \neq 0$, for otherwise we have the contradiction $p_i = m_{f_i} = X$. Thus $a_i = 0$ and so $p_i = X^2 + b_i^2$ where $b_i \neq 0$. $\diamond$

10.5 Corollary *If f is skew-adjoint then the minimum polynomial of f is given as follows :*
(1) *if $f = 0$ then $m_f = X$;*

(2) *if f is invertible then $m_f = \prod_{i=1}^{k} (X^2 + c_i^2)$ for distinct non-zero real numbers $c_1, \ldots, c_k$;*

(3) *if f is neither 0 nor invertible then $m_f = X \prod_{i=1}^{k} (X^2 + c_i^2)$ for distinct non-zero real numbers $c_1, \ldots, c_k$.*

Proof This is immediate from 10.3 and 10.4 on recalling that f is invertible if and only if the constant term in m_f is non-zero. $\Diamond$

We now observe how orthogonality comes into the picture.

10.6 Theorem *If V is a non-zero finite-dimensional real inner product space and if $f : V \to V$ is skew-adjoint then the primary components of f are pairwise orthogonal.*

Proof Let V_i, V_j be primary components of f with $i \neq j$. If, with the usual notation, f_i, f_j are the mappings induced on V_i, V_j suppose first that $m_{f_i} = X^2 + c_i^2$ and $m_{f_j} = X^2 + c_j^2$ where c_i, c_j are not zero and $c_i^2 \neq c_j^2$. Then for $x_i \in V_i$ and $x_j \in V_j$ we have

$$\begin{aligned} 0 &= \langle (f_i^2 + c_i^2 \operatorname{id}_{V_i} \,|\, x_j \rangle \\ &= \langle f^2(x_i) \,|\, x_j \rangle + c_i^2 \langle x_i \,|\, x_j \rangle \\ &= \langle f(x_i) \,|\, -f(x_j) \rangle + c_i^2 \langle x_i \,|\, x_j \rangle \\ &= \langle x_i \,|\, f^2(x_j) \rangle + c_i^2 \langle x_i \,|\, x_j \rangle \\ &= \langle x_i \,|\, -c_j^2 x_j \rangle + c_i^2 \langle x_i \,|\, x_j \rangle \\ &= (c_i^2 - c_j^2) \langle x_i \,|\, x_j \rangle. \end{aligned}$$

Since $c_i^2 \neq c_j^2$ it follows that $\langle x_i \,|\, x_j \rangle = 0$.

Suppose now that $m_{f_i} = X^2 + c_i^2$ with $c_i \neq 0$ and $m_{f_j} = X$. Replacing $f^2(x_j)$ by 0_V in the above string of equalities we obtain $0 = c_i^2 \langle x_i \,|\, x_j \rangle$ whence again $\langle x_i \,|\, x_j \rangle = 0$. $\Diamond$

One further result and we can establish our main theorem on skew-adjoint mappings.

10.7 Theorem *Let V be a non-zero finite-dimensional inner product space and let W be a subspace of V. Then W is f-invariant if and only if $W^\perp$ is f^*-invariant.*

Proof Let W be f-invariant. Since $V = W \oplus W^\perp$ we have

$$(\forall x \in W)(\forall y \in W^\perp) \qquad \langle x \mid f^*(y) \rangle = \langle f(x) \mid y \rangle = 0$$

whence $f^*(y) \in W^\perp$ for all $y \in W^\perp$, so $W^\perp$ is f^*-invariant. Applying this observation again, we obtain the converse; for if $W^\perp$ is f^*-invariant then $W = W^{\perp\perp}$ is $f^{**} = f$-invariant. $\diamond$

10.8 Theorem *Let V be a non-zero finite-dimensional real inner product space and let $f : V \to V$ be a skew-adjoint mapping with $m_f = X^2 + b^2$ where $b \neq 0$. Then $\dim V$ is even and there is an ordered orthonormal basis of V with respect to which the matrix of f is of the form*

$$M(b) = \begin{bmatrix} 0 & -b & & & & & \\ b & 0 & & & & & \\ & & 0 & -b & & & \\ & & b & 0 & & & \\ & & & & \ddots & & \\ & & & & & 0 & -b \\ & & & & & b & 0 \end{bmatrix}.$$

Proof We begin by showing that $\dim V$ is even and that V is an orthogonal direct sum of f-cyclic subspaces each of dimension 2. For this purpose, let y be a non-zero element of V. Observe that $f(y) \neq \lambda y$ for any λ; otherwise, since $f^2(y) = -b^2 y$, we would have $\lambda^2 = -b^2$ and hence the contradiction $b = 0$. Let W_1 be the smallest f-invariant subspace containing y. Since $f^2(y) = -b^2 y$ it follows that W_1 is f-cyclic of dimension 2, a cyclic basis for W_1 being $\{y, f(y)\}$. Consider now the decomposition $V = W_1 \oplus W_1^\perp$. This direct sum is orthogonal; for if p is the projection on W_1 parallel to $W_1^\perp$ then $\text{Im}\, p = W_1$ and $\text{Ker}\, p = W_1^\perp$ so p is an orthoprojection. By 10.7, $W_1^\perp$ is f^*-invariant and so, since $f^* = -f$, we see that $W_1^\perp$ is also f-invariant, of dimension $\dim V - 2$. Now let $V_1 = W_1^\perp$ and repeat the argument to obtain an orthogonal direct sum $V_1 = W_2 \oplus W_2^\perp$ of f-invariant subspaces with W_2 f-cyclic of dimension 2. Continuing in this manner, we note that it is not possible in the final such decomposition to have $\dim W_n^\perp = 1$. For, if this were so then $W_n^\perp$ would have a singleton basis $\{z\}$ whence $f(z) \notin W_n^\perp$, a contradiction. Thus $W_n^\perp$ also has a basis of the form $\{z, f(z)\}$, so is likewise f-cyclic of dimension 2. It follows that $\dim V$ is even.

We now construct an orthonormal basis for each of the f-cyclic subspaces W_i. Consider the basis $\{y_i, f(y_i)\}$. Since

$$\|f(y_i)\|^2 = \langle f(y_i) \,|\, -f^*(y_i)\rangle = -\langle f^2(y_i)\,|\,y_i\rangle = b^2\|y_i\|^2$$

it follows by applying the Gram-Schmidt process that an orthonormal basis for W_i is

$$B_i = \left\{\frac{y_i}{\|y_i\|}, \frac{f(y_i)}{b\|y_i\|}\right\}.$$

The matrix of f_i relative to B_i is then readily seen to be

$$\begin{bmatrix} 0 & -b \\ b & 0 \end{bmatrix}.$$

Pasting together such bases, we obtain an orthonormal basis of V with respect to which the matrix of f is of the form stated. $\diamondsuit$

10.9 Corollary *If V is a non-zero finite-dimensional real inner product space and if $f : V \to V$ is skew-adjoint then there is an ordered orthonormal basis of V with respect to which the matrix of f is of the form*

$$\begin{bmatrix} M_1 & & \\ & \ddots & \\ & & M_k \end{bmatrix}$$

where each M_i is either 0 or as described in 10.8.

Proof Combine 10.5, 10.6 and 10.8. $\diamondsuit$

10.10 Corollary *A real square matrix is skew-symmetric if and only if it is orthogonally similar to a matrix of the form given in 10.9.* $\diamondsuit$

We now turn to the general problem of normal linear mappings on a real inner product space. Recall from 10.1 that such a mapping can be expressed uniquely in the form $g + h$ where g is self-adjoint and h is skew-adjoint. Moreover, by 8.11, g is ortho-diagonalizable.

10.11 Theorem *Let V be a non-zero finite-dimensional real inner product space and let $f : V \to V$ be a normal linear mapping whose minimum polynomial is $m_f = X^2 - 2aX + a^2 + b^2$ where $b \neq 0$. If g, h are the self-adjoint and skew-adjoint parts of f then*

(1) *h is invertible;*
(2) *$m_g = X - a$;*
(3) *$m_h = X^2 + b^2$.*

Proof (1) Suppose, by way of obtaining a contradiction, that $\operatorname{Ker} h \neq \{0_V\}$. Since f is normal we have $g \circ h = h \circ g$ by 10.1, from which we see that $\operatorname{Ker} h$ is g-invariant. Since $f = g + h$, the restriction of f to $\operatorname{Ker} h$ coincides with that of g. As $\operatorname{Ker} h$ is g-invariant, we can therefore define a linear mapping $f' : \operatorname{Ker} h \to \operatorname{Ker} h$ by the prescription $f'(x) = f(x) = g(x)$; and since g is self-adjoint so is f'. By 8.11, f' is then ortho-diagonalizable, and so its minimum polynomial is a product of distinct linear factors. But $m_{f'}$ must divide m_f which, by hypothesis, is irreducible. This contradiction therefore gives $\operatorname{Ker} h = \{0_V\}$ whence h is invertible.

(2) Since $f = g + h$ with $g^\star = g$ and $h^\star = -h$ we have $f^\star = g - h$. As $f^2 - 2af + (a^2 + b^2)\operatorname{id}_V = 0$ we have $(f^\star)^2 - 2af^\star + (a^2 + b^2)\operatorname{id}_V = 0$ and consequently $f^2 - (f^\star)^2 = 2a(f - f^\star) = 4ah$. Thus, since $f \circ f^\star = f^\star \circ f$, we see that

$$g \circ h = \tfrac{1}{2}(f + f^\star) \circ \tfrac{1}{2}(f - f^\star) = \tfrac{1}{4}[f^2 - (f^\star)^2] = ah$$

and so $(g - a\operatorname{id}_V) \circ h = 0$. Since h is invertible by (1) we then have that $g - a\operatorname{id}_V = 0$, whence $m_g = X - a$.

(3) Since $f - h = g = a\operatorname{id}_V$ we have $f = h + a\operatorname{id}_V$ and so

$$\begin{aligned} 0 &= f^2 - 2af + (a^2 + b^2)\operatorname{id}_V \\ &= (h + a\operatorname{id}_V)^2 - 2a(h + a\operatorname{id}_V) + (a^2 + b^2)\operatorname{id}_V \\ &= h^2 + b^2\operatorname{id}_V .\end{aligned}$$

Now h is skew-adjoint and, by (1), is invertible. It now follows by 10.5 that $m_h = X^2 + b^2$. ◇

10.12 Theorem *If V is a non-zero finite-dimensional real inner product space and if $f : V \to V$ is normal then the primary components of f are pairwise orthogonal.*

Proof By 10.3 the minimum polynomial of f has the general form
$$m_f = (X - a_0) \prod_{i=1}^{k} (X^2 - 2a_i X + a_i^2 + b_i^2)$$
where each $b_i \neq 0$. The primary components of f are therefore given by $V_0 = \mathrm{Ker}(f - a_0 \, \mathrm{id}_V)$ and
$$(i = 1, \ldots, k) \qquad V_i = \mathrm{Ker}[f^2 - 2a_i f + (a_i^2 + b_i^2) \, \mathrm{id}_V].$$

Also, the induced mapping f_i on V_i is normal, with minimum polynomial $X - a_0$ if $i = 0$ and $X^2 - 2a_i X + a_i^2 + b_i^2$ otherwise. Now $f_i = g_i + h_i$ where g_i is self-adjoint and h_i is skew-adjoint. Moreover, g_i, h_i coincide with the mappings induced on V_i by g and h. To see this, let these mappings be g', h' respectively. Then for every $x \in V_i$ we have $g_i(x) + h_i(x) = f_i(x) = f(x) = g(x) + h(x) = g'(x) + h'(x)$ and so $g_i - g' = h' - h_i$. Since the left hand side is self-adjoint and the right hand side is skew-adjoint we deduce that $g_i = g'$ and $h_i = h'$.

Suppose now that $i, j > 0$ with $i \neq j$. Then the minimum polynomials of f_i, f_j are $X^2 - 2a_i X + a_i^2 + b_i^2$ and $X^2 - 2a_j X + a_j^2 + b_j^2$ where either $a_i \neq a_j$ or $b_i^2 \neq b_j^2$. By 10.11, we have $m_{g_i} = X - a_i, m_{g_j} = X - a_j$ and $m_{h_i} = X^2 + b_i^2, m_{h_j} = X^2 + b_j^2$. Given $x_i \in V_i$ and $x_j \in V_j$ we therefore have

$$\begin{aligned}
0 = \langle (h_i^2 + b_i^2 \, \mathrm{id}_{V_i})(x_i) \,|\, x_j \rangle &= \langle h^2(x_i) \,|\, x_j \rangle + b_i^2 \langle x_i \,|\, x_j \rangle \\
&= \langle x_i \,|\, h^2(x_j) \rangle + b_i^2 \langle x_i \,|\, x_j \rangle \\
&= \langle x_i \,|\, h_j^2(x_j) \rangle + b_i^2 \langle x_i \,|\, x_j \rangle \\
&= -b_j^2 \langle x_i \,|\, x_j \rangle + b_i^2 \langle x_i \,|\, x_j \rangle \\
&= (b_i^2 - b_j^2) \langle x_i \,|\, x_j \rangle,
\end{aligned}$$

so that in the case where $b_i^2 \neq b_j^2$ we have $\langle x_i \,|\, x_j \rangle = 0$. Likewise,

$$\begin{aligned}
0 = \langle (g_i - a_i \, \mathrm{id}_{V_i})(x_i) \,|\, x_j \rangle &= \langle g(x_i) \,|\, x_j \rangle - a_i \langle x_i \,|\, x_j \rangle \\
&= \langle x_i \,|\, g(x_j) \rangle - a_i \langle x_i \,|\, x_j \rangle \\
&= \langle x_i \,|\, g_j(x_j) \rangle - a_i \langle x_i \,|\, x_j \rangle \\
&= a_j \langle x_i \,|\, x_j \rangle - a_i \langle x_i \,|\, x_j \rangle \\
&= (a_j - a_i) \langle x_i \,|\, x_j \rangle
\end{aligned}$$

so that in the case where $a_i \neq a_j$ we have $\langle x_i \,|\, x_j \rangle = 0$. We thus see that $V_1, \ldots, V_k$ are pairwise othogonal. That V_0 is also orthogonal to each V_i for $i \geq 1$ follows from the above strings of equalities on taking $j = 0$ and using the fact that $f_0 = a_0 \operatorname{id}_{V_0}$ is self-adjoint and consequently $g_0 = f_0$ and $h_0 = 0$. $\diamond$

We can now establish the main result.

10.13 Theorem *If V is a non-zero finite-dimensional real inner product space and if $f : V \to V$ is a normal linear mapping then there is an ordered orthonormal basis of V relative to which the matrix of f is of the form*

$$\begin{bmatrix} A_1 & & & \\ & A_2 & & \\ & & \ddots & \\ & & & A_k \end{bmatrix}$$

where each A_i is either a 1×1 matrix or a 2×2 matrix of the form

$$\begin{bmatrix} \alpha & -\beta \\ \beta & \alpha \end{bmatrix}$$

in which $\beta \neq 0$.

Proof With the same notation as above, let

$$m_f = (X - a_0) \prod_{i=1}^{k} (X^2 - 2a_i X + a_i^2 + b_i^2)$$

and let the primary components of f be V_i for $i = 0, \ldots, k$. Then $m_{f_i} = X - a_0$ if $i = 0$ and $m_{f_i} = X^2 - 2a_i X + a_i^2 + b_i^2$ otherwise.

Given any V_i with $i \neq 0$ we have $f_i = g_i + h_i$ where the self-adjoint part g_i has minimum polynomial $X - a_i$ and the skew-adjoint part h_i has minimum polynomial $X^2 + b_i^2$. Now h_i is skew-adjoint and so, by 10.8, there is an ordered orthonormal basis B_i of V_i with respect to which the matrix of h_i is

$$M(b_i) = \begin{bmatrix} 0 & -b_i & & & & & \\ b_i & 0 & & & & & \\ & & 0 & -b_i & & & \\ & & b_i & 0 & & & \\ & & & & \ddots & & \\ & & & & & 0 & -b_i \\ & & & & & b_i & 0 \end{bmatrix}.$$

Since the minimum polynomial of g_i is $X - a_i$ we have $g_i(x) = a_i x$ for every $x \in B_i$ and so the matrix of g_i relative to B_i is the diagonal matrix all of whose diagonal entries are a_i. It now follows that the matrix of $f_i = g_i + h_i$ relative to B_i is

$$M(a_i, b_i) = \begin{bmatrix} a_i & -b_i & & & & & \\ b_i & a_i & & & & & \\ & & a_i & -b_i & & & \\ & & b_i & a_i & & & \\ & & & & \ddots & & \\ & & & & & a_i & -b_i \\ & & & & & b_i & a_i \end{bmatrix}.$$

In the case where $i = 0$, we have $f_0 = a_0 \, \text{id}_{V_0}$ so f_0 is self-adjoint. By 8.11, there is an ordered orthonormal basis of V_0 with respect to which the matrix of f_0 is diagonal.

Now by 10.12 the primary components V_i are pairwise orthogonal. Pasting together the ordered orthonormal bases in question, we then obtain an ordered orthonormal basis of V relative to which the matrix of f is of the form stated. $\diamond$

10.14 Corollary *A real square matrix is normal if and only if it is orthogonally similar to a matrix of the form described in 10.13.* $\diamond$

Our labours produce a bonus : an orthogonal linear mapping f is such that f^{-1} exists and equals f^*, and so is in particular normal. We can therefore deduce from the above a canonical form for orthogonal mappings and matrices.

10.15 Theorem *If V is a non-zero finite-dimensional real inner product space and $f : V \to V$ is an orthogonal linear mapping then there is an ordered orthonormal basis of V with respect to which the matrix of f is of the form*

$$\begin{bmatrix} I_m & & & & & \\ & -I_p & & & & \\ & & P_1 & & & \\ & & & P_2 & & \\ & & & & \ddots & \\ & & & & & P_k \end{bmatrix}$$

in which each P_i is a 2×2 matrix of the form

$$\begin{bmatrix} \alpha & -\beta \\ \beta & \alpha \end{bmatrix}$$

where $\beta \neq 0$ and $\alpha^2 + \beta^2 = 1$.

Proof With the same notation as in 10.13, we have that the matrix $M(a_i, b_i)$, which represents f_i relative to the ordered basis B_i, is an orthogonal matrix (since f_i is orthogonal). Multiplying this matrix by its transpose, we obtain an identity matrix and, equating entries, we see that $a_i^2 + b_i^2 = 1$. As for the primary component V_0, the matrix of f_0 is diagonal. Since the square of this diagonal matrix is an identity matrix, its entries must be ± 1. We can now rearrange the basis to see that the matrix of f has the form described. $\diamond$

Example If $f : \mathbb{R}^3 \to \mathbb{R}^3$ is orthogonal then f is called a *rotation* if $\det A = 1$ for any matrix A that represents f. If f is a rotation then there is an ordered orthonormal basis of $\mathbb{R}^3$ with respect to which the matrix of f is

$$\begin{bmatrix} 1 & 0 & 0 \\ 0 & \cos\vartheta & -\sin\vartheta \\ 0 & \sin\vartheta & \cos\vartheta \end{bmatrix}$$

for some real number ϑ.

Index

algebra, 1
annihilator, 47,67
adjoint, 80,85

Bessel's inequality, 73
bidual, 63
bilinear form, 99
bitranspose, 66
block diagonal form, 15

Cayley-Hamilton theorem, 2
Cauchy-Schwarz inequality, 71
characteristic polynomial, 2
classical canonical matrix, 56
classical p-matrix, 56
companion matrix, 49
complex inner product space, 69
conjugate isomorphism, 78
conjugate transformation, 78
coordinate form, 60
cyclic basis, 49
cyclic decomposition, 49
cyclic subspace, 49

diagonalizable, 20
direct sum, 8
distance, 71
dot product, 70
dual space, 58

elementary divisor, 53
elementary Jordan matrix, 33
eigenvalue, 5
eigenvector, 5

Fourier coefficients, 77

Gram matrix, 98
Gram-Schmidt process, 75

Hilbert space, 72

idempotent, 10
index, 31
inner product space, 69
invariant subspace, 13

Jordan basis, 39
Jordan block, 33
Jordan canonical matrix, 37
Jordan decomposition, 28

Lagrange polynomial
linear form, 58
linear functional, 58

mimimum polynomial, 3

nilpotent, 22
normalising, 73

ortho-diagonalizable, 88
orthogonal, 72
orthogonal complement, 82
orthogonal direct sum, 87
orthogonally similar, 86
orthonormal, 72
orthonormal basis, 74
ortho-projection, 87

Parseval's identity, 77
positive, 96
positive definite, 96, 106
primary decomposition, 15
projection, 10

quadratic form, 101
quotient space, 44

rational canonical matrix, 53
real inner product space, 69

scalar product, 70
signature, 105
simultaneously diagonalizable, 21
skew-adjoint, 108
spectral resolution, 90
square summable, 72
sum of subspaces, 7
Sylvester's theorem, 105
symmetric bilinear form, 101

triangular form, 24
triangle inequality, 71

unitarily similar, 86
unitary, 86